NASA/SP-2011-3422
Version 1.0

NASA
Risk Management
Handbook

National Aeronautics and Space Administration
NASA Headquarters
Washington, D.C. 20546

November 2011

Published by Books Express Publishing
Copyright © Books Express, 2012
ISBN 978-1-78266-137-5

Books Express publications are available from all good retail and online booksellers. For
publishing proposals and direct ordering please contact us at: info@books-express.com

NASA STI Program ... in Profile

Since its founding, NASA has been dedicated to the advancement of aeronautics and space science. The NASA scientific and technical information (STI) program plays a key part in helping NASA maintain this important role.

The NASA STI program operates under the auspices of the Agency Chief Information Officer. It collects, organizes, provides for archiving, and disseminates NASA's STI. The NASA STI program provides access to the NASA Aeronautics and Space Database and its public interface, the NASA Technical Report Server, thus providing one of the largest collections of aeronautical and space science STI in the world. Results are published in both non-NASA channels and by NASA in the NASA STI Report Series, which includes the following report types:

TECHNICAL PUBLICATION. Reports of completed research or a major significant phase of research that present the results of NASA Programs and include extensive data or theoretical analysis. Includes compilations of significant scientific and technical data and information deemed to be of continuing reference value. NASA counterpart of peer-reviewed formal professional papers but has less stringent limitations on manuscript length and extent of graphic presentations.

TECHNICAL MEMORANDUM. Scientific and technical findings that are preliminary or of specialized interest, e.g., quick release reports, working papers, and bibliographies that contain minimal annotation. Does not contain extensive analysis.

CONTRACTOR REPORT. Scientific and technical findings by NASA-sponsored contractors and grantees.

CONFERENCE PUBLICATION. Collected papers from scientific and technical conferences, symposia, seminars, or other meetings sponsored or co-sponsored by NASA.

SPECIAL PUBLICATION. Scientific, technical, or historical information from NASA programs, projects, and missions, often concerned with subjects having substantial public interest.

TECHNICAL TRANSLATION. English-language translations of foreign scientific and technical material pertinent to NASA's mission.

Specialized services also include creating custom thesauri, building customized databases, and organizing and publishing research results.

For more information about the NASA STI program, see the following:

Access the NASA STI program home page at *http://www.sti.nasa.gov*

E-mail your question via the Internet to help@sti.nasa.gov

Fax your question to the NASA STI Help Desk at 443-757-5803

Phone the NASA STI Help Desk at 443-757-5802

Write to:

NASA STI Help Desk
NASA Center for AeroSpace Information
7115 Standard Drive
Hanover, MD 21076-1320

ACKNOWLEDGMENTS

The project manager and the authors express their gratitude to NASA Office of Safety and Mission Assurance (OSMA) management (Mr. Bryan O'Connor, former Chief of OSMA; Mr. Terrence Wilcutt, Chief of OSMA; Mr. Wilson Harkins, Deputy Chief of OSMA; and Mr. Thomas Whitmeyer, Director of Mission Support Division) for their support and encouragement in developing this document. The development effort leading to this document was conducted in stages, and was supported by the individuals listed alphabetically below, who each brought unique experience and insights to the development.[1]

AUTHORS:

Dr. Homayoon Dezfuli (Project Manager)	NASA System Safety Technical Fellow, NASA Headquarters
Dr. Allan Benjamin	Information Systems Laboratories
Mr. Christopher Everett	Information Systems Laboratories
Mr. Gaspare Maggio	Information Systems Laboratories
Dr. Michael Stamatelatos	Director of Safety and Assurance Requirements Division, NASA Headquarters
Dr. Robert Youngblood	Idaho National Laboratory

CONTRIBUTING AUTHORS:

Dr. Sergio Guarro	The Aerospace Corporation
Dr. Peter Rutledge	Quality Assurance & Risk Management Services
Mr. James Sherrard	Information Systems Laboratories
Dr. Curtis Smith	Idaho National Laboratory
Dr. Rodney Williams	Information Systems Laboratories

REVIEWERS:

This development benefited from review comments provided on the initial draft by many individuals. The authors wish to specifically thank the following individuals:

Dr. Robert Abelson	NASA Jet Propulsion Laboratory
Dr. Timothy Barth	NASA Kennedy Space Center
Mr. John Chiorini	Center for Systems Management
Mr. Chester Everline	Jet Propulsion Laboratory
Mr. Louis Fussell	Futron Corporation
Dr. Frank Groen	NASA Headquarters
Mr. David Lengyel	NASA Headquarters
Dr. Robert Mulvihill	Quality Assurance & Risk Management Services
Ms. Sylvia Plants	Science Applications International Corporation

---

[1] Affiliations are as of the time of contribution to the development effort.

Mr. William Powell	NASA Marshall Space Flight Center
Mr. David Pye	Perot Systems
Dr. James Rose	Jet Propulsion Laboratory
Dr. Fayssal Safie	NASA Marshall Space Flight Center
Dr. Nathan Siu	U. S. Nuclear Regulatory Commission
Dr. Clayton Smith	Applied Physics Laboratory
Ms. Sharon Thomas	NASA Johnson Space Center
Ms. Ellen Stigberg	NASA Headquarters
Dr. William Vesely	NASA Headquarters
Mr. Tracy Wrigley	Bastion Technologies, Inc.
Dr. Thomas Zang	NASA Langley Research Center

TABLE OF CONTENTS

LIST OF FIGURES

LIST OF TABLES

NASA RISK MANAGEMENT HANDBOOK

Preface

In some form, risk management (RM) has always been an integral part of virtually every challenging human endeavor. A formal and, at that time, qualitative RM process known as Continuous Risk Management (CRM) was introduced to NASA in the latter half of the 1990s. More rigorous quantitative RM processes including Risk-Informed Decision Making (RIDM) and an enhanced version of CRM have only recently been developed for implementation as an integral part of systems engineering at NASA. While there will probably always be vigorous debate over the details of what comprises the best approach to managing risk, few will disagree that effective risk management is critical to program and project success and affordability.

Since their introduction and until recently, NASA RM processes have been based on CRM, which stresses the management of risk during the Implementation phase of the NASA Program/Project Life Cycle. In December of 2008, NASA issued NPR 8000.4A [1], which introduced RIDM as a complementary process to CRM that is concerned with analysis of important and/or direction-setting decisions. In the past, RM was considered equivalent to CRM; now, RM is defined as comprising both CRM and RIDM.

In April 2010, NASA issued NASA/SP-2010-576, the NASA Risk-Informed Decision Making Handbook. This handbook introduced RIDM as the front-end of the RM process, described the details of how RIDM is conducted, and ended with a description of how the results of RIDM transition to and set the stage for CRM, the final portion of the RM process. The RIDM Handbook did not proceed to describe CRM, as the development of an enhanced version of CRM was still a work in progress in 2010.

Now this handbook addresses the entirety of the NASA RM process, including both RIDM and CRM. Beginning with and facilitated by RIDM, decisions made during the course of a program ultimately "burn in" the risk that must be managed during the life cycle of the program (primarily during the development portion of the life cycle) using CRM processes to ensure progress towards the program's goal. RIDM helps to ensure that decisions between alternatives are made with an awareness of the risks associated with each, thereby helping to prevent late design changes, which can be key drivers of risk, cost overruns, schedule delays, and cancellation. Most project cost-saving opportunities occur in the definition, planning, and early design phases of a project [2]. After being initialized by the results of RIDM, CRM is used to manage the aggregate risk that threatens the achievement of performance requirements. It does so based on a given set of performance requirements and decision maker risk tolerance levels, analyzing identified risk scenarios with possible mitigations and with follow-up monitoring and communications; by maintaining current and modifying as needed the RIDM risk models; by identifying new risks as they arise and including in the models those risks that were not considered discriminators in RIDM; by documenting individual risks in the form of risk statements with accompanying descriptive narratives for complete understanding; by analyzing departure events; by estimating aggregate risk and the criticality of individual risks; by developing risk scenarios leading to the analysis of pivotal events and the identification of risk drivers; by developing aggregate risk models for performance requirements; by identifying risk mitigation options and new risks that may arise from their implementation; by tracking and controlling the effectiveness of mitigations; and finally, by communicating and documenting all risk information necessary to an effective RM process.

The RIDM process described in this document attempts to respond to some of the primary issues that have derailed programs in the past, namely: 1) the "mismatch" between stakeholder expectations and the "true" resources required to address the risks to achieve those expectations; 2) the miscomprehension of the risk that a decision maker is accepting when making commitments to stakeholders; and 3) the miscommunication in considering the respective risks associated with competing alternatives.

The CRM process described herein is an enhanced version of NASA's traditional CRM paradigm. While it maintains the traditional core elements of CRM as we have known them in the past, it builds upon the solid foundation of quantitative parameters and data made possible by the RIDM front-end to the RM process. This approach fundamentally changes the focus from qualitative assessments to quantitative analyses, from the management of individual risks to the management of aggregate risk, and from eliminating or reducing the impact of single unwanted events to the management of risk drivers. This quantification allows managers to discover the drivers of the total risk and find the interactions and dependencies among their causes, mitigations, and impacts across all parts of the program's organization. In addition, quantification supports the optimization of constrained resources, leading to greater affordability. Armed with a set of performance requirements and knowledge of the decision maker's risk tolerance, CRM is used to manage the individual risks that collectively contribute to the aggregate risk of not meeting program/project performance requirements and goals.

This handbook is primarily written for systems engineers, risk managers, and risk analysts assigned to apply the requirements of NPR 8000.4A. However, program managers of NASA programs and projects can also get a sense of the value added of the process by reading the RIDM and CRM overview sections. These sections are designed to provide concise descriptions of RIDM and CRM and to highlight key areas of the processes.

The RM methodology introduced by this handbook is part of a systems engineering process which emphasizes the proper use of risk analysis in its broadest sense to make risk-informed decisions that impact the mission execution domains of safety, technical, cost, and schedule. In future versions of this handbook, the RM principles discussed here will be updated in an evolutionary manner and expanded to address operational procedures, procurement, strategic planning, and institutional RM as experience is gained in the field. Additionally, technical appendices will be developed and added to provide tools and templates for implementation of the RM process.

This handbook has been informed by many other guidance efforts underway at NASA, including the NASA Systems Engineering Handbook (NASA/SP 2007 6105 Rev. 1), the 2008 NASA Cost Estimating Handbook (NASA CEH 2008), and the NASA Standard for Models and Simulation (NASA STD 7009) to name a few. How these documents relate and interact with the RM Handbook is discussed in subsequent chapters. With this in mind, this handbook could be seen as a complement to those efforts in order to help ensure programmatic success and affordability. In fact, the RM methodology has been formulated to complement, but not duplicate, the guidance in those documents. Taken together, the overall guidance is meant to maximize program/project success and affordability by providing: 1) systematic and well thought out processes for conducting the discipline processes as well as integrating them into a formal risk analysis framework and communicating those results to a decision maker so that he or she can make the best informed decisions possible, and 2) a systematic and rigorous process that manages individual and aggregate risks in order to meet program/project performance requirements and goals within levels of risk considered tolerable to the involved decision maker.

Although formal decision analysis methods are now highly developed for unitary decision-makers, it is still a significant challenge to apply these methods in a practical way within a complex organizational hierarchy having its own highly developed program management policies and practices. This handbook is a step towards meeting that challenge for NASA but is certainly not the final step in realizing the proper balance between formalism and practicality. Therefore, efforts will continue to ensure that the methods in this document are properly integrated and updated as necessary, to provide value to the program and project management processes at NASA.

While the RM process described in this handbook currently focuses mainly on risks to the achievement of numerical mission performance requirements associated with the safety, technical, cost, and schedule mission execution domains, future revisions of this document will address institutional, enterprise, and Agency-wide strategic risks. In the meantime, it should be noted that the techniques presented here may well be applicable to these latter types of risks, even now.

Finally, it is important to point out that *this handbook is not a prescription for how to do risk management*. Rather, it is guidance on how RM can be done in an integrated framework that flows through a logical and carefully thought-out sequence of related activities that can and should always be tailored to the situation at hand.

Homayoon Dezfuli, Ph.D.
Project Manager, NASA Headquarters
November 2011

NASA RISK MANAGEMENT HANDBOOK

1   INTRODUCTION

1.1   Purpose

The purpose of this handbook is to provide guidance for implementing the Risk Management (RM) requirements of NASA Procedural Requirements (NPR) document NPR 8000.4A, Agency Risk Management Procedural Requirements [1], with a specific focus on programs and projects, and applying to each level of the NASA organizational hierarchy as requirements flow down.

This handbook supports RM application within the NASA systems engineering process, and is a complement to the guidance contained in NASA/SP-2007-6105, NASA Systems Engineering Handbook [2]. Specifically, this handbook provides guidance that is applicable to the common technical processes of *Technical Risk Management* and *Decision Analysis* established by NPR 7123.1A, NASA Systems Engineering Process and Requirements [3]. These processes are part of the "Systems Engineering Engine" (Figure 1) that is used to drive the development of the system and associated work products to satisfy stakeholder expectations in all mission execution domains, including safety, technical, cost, and schedule.

Like NPR 7123.1A, NPR 8000.4A is a discipline-oriented NPR that intersects with product-oriented NPRs such as NPR 7120.5D, NASA Space Flight Program and Project Management Requirements [4]; NPR 7120.7, NASA Information Technology and Institutional Infrastructure Program and Project Management Requirements [5]; and NPR 7120.8, NASA Research and Technology Program and Project Management Requirements [6]. In much the same way that the NASA Systems Engineering Handbook is intended to provide guidance on the implementation of NPR 7123.1A, this handbook is intended to provide guidance on the implementation of NPR 8000.4A.

1.2   Scope and Depth

This handbook provides guidance for conducting RM in the context of NASA program and project life cycles, which produce derived requirements in accordance with existing systems engineering practices that flow down through the NASA organizational hierarchy. The guidance in this handbook is not meant to be prescriptive. Instead, it is meant to be general enough, and contain a sufficient diversity of examples, to enable the reader to adapt the methods as needed to the particular risk management issues that he or she faces. The handbook highlights major issues to consider when managing programs and projects in the presence of potentially significant uncertainty, so that the user is better able to recognize and avoid pitfalls that might otherwise be experienced.

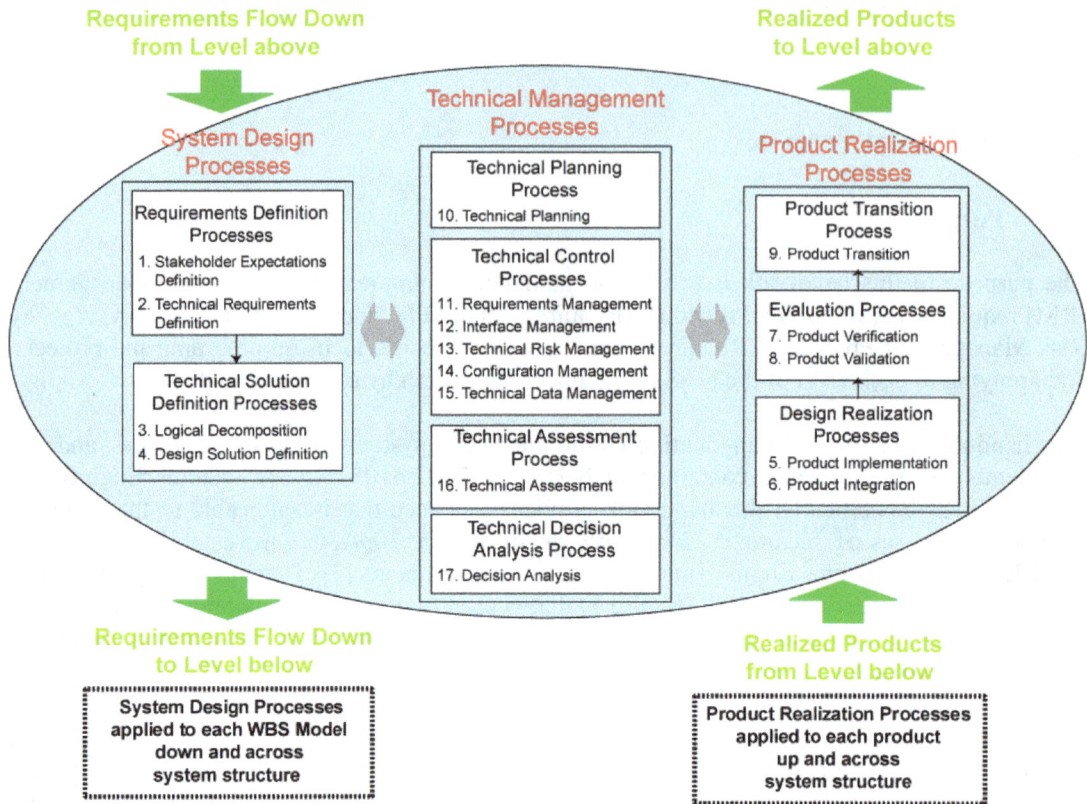

Figure 1. Systems Engineering Engine

Examples are provided throughout the handbook to illustrate the application of RM methods to specific issues of the type that are routinely encountered in NASA programs and projects. An example notional planetary mission is postulated and used throughout the document as a basis for illustrating the execution of the various process steps that constitute risk management in a NASA context ("yellow boxes"). In addition, key terms and concepts are defined throughout the document ("blue boxes").

Where applicable, guidance is also given on the spectrum of techniques that are appropriate to use, given the spectrum of circumstances under which risks are managed, ranging from narrow-scope risk management at the hardware component level that must be accomplished using a minimum of time and resources, to broad-scope risk management involving multiple organizations upon which significant resources may be brought to bear. The fact that new techniques are discussed is not intended to automatically imply that a whole new set of analyses is needed. Rather, the risk analyses should take maximum advantage of existing activities, while also influencing them as needed in order to produce results that address objectives, at an appropriate level of rigor to support robust decision making. In all cases, the goal is to apply a level of effort to the task of risk management that provides assurance that objectives are met.

1.3   Background

NPR 8000.4A provides the requirements for risk management for the Agency, its institutions, and its programs and projects as required by NASA Policy Directive (NPD) 1000.5, Policy for NASA Acquisition [7]; NPD 7120.4C, Program/Project Management [8]; and NPD 8700.1, NASA Policy for Safety and Mission Success [9].

As discussed in NPR 8000.4A, risk is the potential for performance shortfalls, which may be realized in the future, with respect to achieving explicitly established and stated performance requirements. The performance shortfalls may be related to institutional support for mission execution[2] or related to any one or more of the following mission execution domains:

- Safety

- Technical

- Cost

- Schedule

In order to foster proactive risk management, NPR 8000.4A integrates two complementary processes, Risk-Informed Decision Making (RIDM) and Continuous Risk Management (CRM), into a single coherent framework. The RIDM process addresses the risk-informed selection of decision alternatives to assure effective approaches to achieving objectives, and the CRM process addresses implementation of the selected alternative to assure that requirements are met. These two aspects work together to assure effective risk management as NASA programs and projects are conceived, developed, and executed. Figure 2 illustrates the concept.

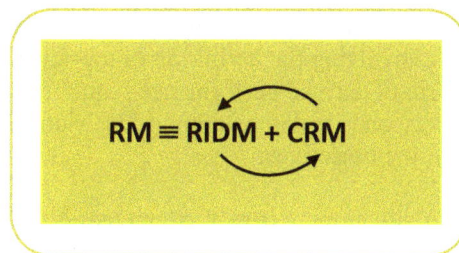

Figure 2. Risk Management as the Interaction of Risk-Informed Decision Making and Continuous Risk Management

Within the NASA organizational hierarchy, high-level objectives, in the form of NASA Strategic Goals, flow down in the form of progressively more detailed performance requirements, whose satisfaction assures that the objectives are met. Each organizational unit within NASA negotiates

---

[2] For the purposes of this version of the handbook, performance shortfalls related to institutional support for mission execution are subsumed under the affected mission execution domains of the program or project under consideration. More explicit consideration of institutional risks will be provided in future versions of this handbook.

with the unit(s) at the next lower level in the organizational hierarchy a set of objectives, deliverables, performance measures, performance requirements, resources, and schedules that defines the tasks to be performed by the unit(s). Once established, the lower level organizational unit employs CRM to manage its own risks against these specifications, and, as appropriate, reports risks and elevates decisions for managing risks to the next higher level based on predetermined risk thresholds that have been negotiated between the two units. Figure 3 depicts this concept. Invoking the RIDM process in support of key decisions as requirements flow down through the organizational hierarchy assures that objectives remain tied to NASA Strategic Goals while also capturing why a particular path for satisfying those requirements was chosen. Managing risk using the CRM process assures that risk management decisions are informed by their impact on objectives at every level of the NASA hierarchy.

As applied to CRM, risk is characterized as a set of triplets:

- The *scenario(s)* leading to degraded performance with respect to one or more performance measures (e.g., scenarios leading to injury, fatality, destruction of key assets; scenarios leading to exceedance of mass limits; scenarios leading to cost overruns; scenarios leading to schedule slippage).

- The *likelihood(s)* (qualitative or quantitative) of those scenarios.

- The *consequence(s)* (qualitative or quantitative severity of the performance degradation) that would result if those scenarios were to occur.

Uncertainties are included in the evaluation of likelihoods and consequences.

1.4   Applicability of Risk Management

The RM approach presented in this handbook is applicable to processes conducted within a systems engineering framework, involving the definition of top-level objectives, the flowdown of top-level objectives in the form of derived performance requirements, decision making about the best way to meet requirements, and implementation of decisions in order to achieve the requirements and, consequently, the objectives.

The RM approach is applied in situations where fundamental NASA values in the domains of safety and technical accomplishment have to be balanced against programmatic realities in the domains of schedule and cost. Since the process of implementing an RM approach in itself introduces cost to the project, it is essential that the approach be used in a cost-effective manner. To this end, the methods advocated in this handbook rely on a graded approach to analysis, to manage analysis costs. Because analysis cost is optimized using this approach, the savings achieved by resolving risks before they become problems invariably exceeds the cost of implementing the approach.

Figure 3. Flowdown of Performance Requirements (Illustrative)

1.4.1 When is RIDM Invoked?

RIDM is invoked for key decisions such as architecture and design decisions, make-buy decisions, source selection in major procurements, and budget reallocation (allocation of reserves), which typically involve requirements-setting or rebaseling of requirements.

RIDM is invoked in many different venues, based on the systems engineering and other management processes of the implementing organizational unit. These include boards and panels, authority to proceed milestones, safety review boards, risk reviews, engineering design and operations planning decision forums, configuration management processes, and commit-to-flight reviews, among others.

RIDM is applicable throughout the project life cycle whenever trade studies are conducted. The processes for which decision analysis is typically appropriate, per the NASA Systems Engineering Handbook, are also those for which RIDM is typically appropriate. These decisions typically have one or more of the following characteristics:

- High Stakes — High stakes are involved in the decision, such as significant costs, significant potential safety impacts, or the importance of meeting the objectives.

- Complexity — The actual ramifications of alternatives are difficult to understand without detailed analysis.

- Uncertainty — Uncertainty in key inputs creates substantial uncertainty in the outcome of the decision alternatives and points to risks that may need to be managed.

- Multiple Attributes — Greater numbers of attributes cause a greater need for formal analysis.

- Diversity of Stakeholders — Extra attention is warranted to clarify objectives and formulate performance measures when the set of stakeholders reflects a diversity of values, preferences, and perspectives.

Satisfaction of all of these conditions is not a requirement for conducting RIDM. The point is, rather, that the need for RIDM increases as a function of the above conditions.

1.4.2   When is CRM Applied?

CRM is applied towards the achievement of defined performance requirements. In particular, CRM is applied following the invocation of RIDM for key decisions involving requirements-setting or rebaselining of requirements. CRM processes are applicable at any level of the NASA organizational hierarchy where such requirements are defined, and the CRM processes at each such level are focused on achieving the requirements defined at that level.

Within the context of the NASA program/project life cycle, CRM is applicable during implementation, once performance requirements have been defined. In addition, CRM processes are applicable to formulation activities, such as technology development, involving the achievement of specific objectives within defined cost and schedule constraints.

As implied by its name, CRM entails the continuous management of risks to the performance requirements throughout all phases of implementation, from design and manufacture to operations and eventual closeout, to assure that performance expectations are maintained, and that operational experience is assessed for indications of underappreciated risk.

In the event that one or more performance requirements cannot be met with the risk response options that are available to the project, the CRM process can provide both motivation and justification for seeking waivers from having to meet those requirements. That would be the case when the CRM process is able to show that a given requirement is either unnecessary or counterproductive to the success of the mission.

1.5   Overview of the RIDM Process [10]

As specified in NPR 8000.4A, the RIDM process itself consists of the three parts shown in Figure 4. This section provides an overview of the process and an introduction to the concepts and terminology established for its implementation. A detailed exposition of the steps associated with each part of the process can be found in Section 3, The RIDM Process.

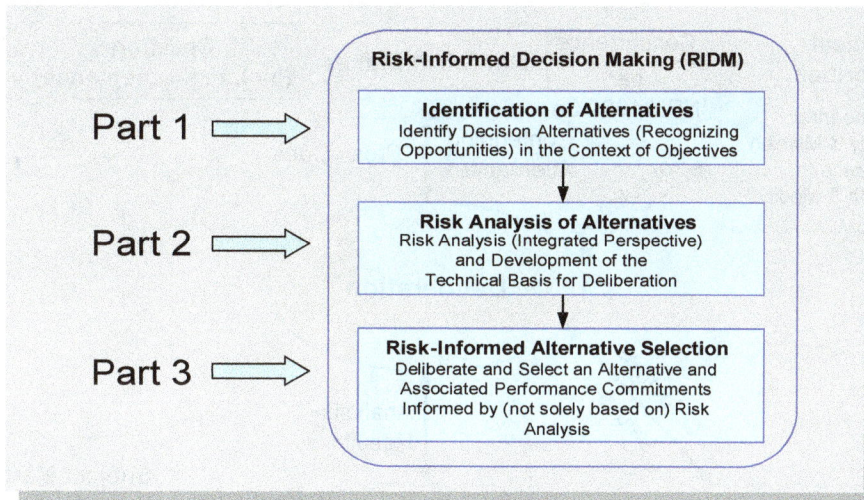

Figure 4. The RIDM Process

Throughout the RIDM process, interactions take place among the ***stakeholders***, the ***risk analysts***, the ***subject matter experts (SMEs)***, the ***Technical Authorities***, and the ***decision-maker*** to ensure that objectives, values, and knowledge are properly integrated and communicated into the deliberations that inform the decision.

Figure 5 notionally illustrates the functional roles and internal interfaces of RIDM. As shown in the figure, it is imperative that the analysts conducting the risk analysis of alternatives incorporate the objectives of the various stakeholders into their analyses. These analyses are performed by, or with the support of, SMEs in the domains spanned by the objectives. The completed risk analyses are deliberated, along with other considerations, and the decision-maker selects a decision alternative for implementation (with the concurrence of the relevant Technical Authorities). The risk associated with the selected decision alternative becomes the central focus of CRM activities, which work to mitigate it during implementation, thus avoiding performance shortfalls in the outcome.

The RIDM process is portrayed in this handbook primarily as a linear sequence of steps, each of which is conducted by individuals in their roles as stakeholders, risk analysts, SMEs, and decision-makers. The linear step-wise approach is used for instructional purposes only. In reality, some portions of the processes may be conducted in parallel, and steps may be iterated upon multiple times before moving to subsequent steps.

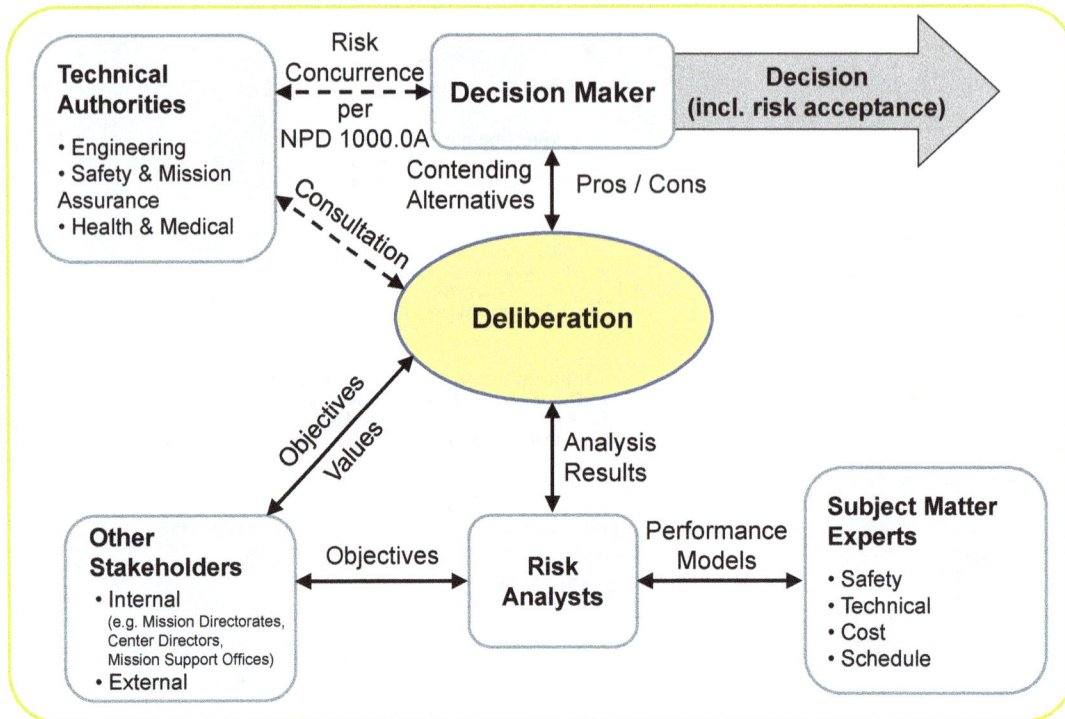

Figure 5. Functional Roles and Information Flow in RIDM (Notional)

RIDM Functional Roles*

Stakeholders - A stakeholder is an individual or organization that is materially affected by the outcome of a decision or deliverable; e.g., Center Directors (CDs), Mission Support Offices (MSOs).

Risk Analysts – A risk analyst is an individual or organization that applies probabilistic methods to the quantification of performance with respect to the mission execution domains of safety, technical, cost, and schedule.

Subject Matter Experts – A subject matter expert is an individual or organization with expertise in one or more topics within the mission execution domains of safety, technical, cost, or schedule.

Technical Authorities – The individuals within the Technical Authority process who are funded independently of a program or project and who have formally delegated Technical Authority traceable to the Administrator. The three organizations who have Technical Authorities are Engineering, Safety and Mission Assurance, and Health and Medical. [11]

Decision-Maker – A decision-maker is an individual with responsibility for decision making within a particular organizational scope.

*Not to be interpreted as official job positions but as functional roles.

In particular, Part 2, Risk Analysis of Alternatives, is internally iterative as analyses are refined to meet decision needs in accordance with a graded approach, and Part 2 is iterative with Part 3, Risk-Informed Alternative Selection, as stakeholders and decision-makers iterate with the risk analysts in order to develop a sufficient technical basis for robust decision making. Additionally, decisions may be made via a series of downselects, each of which is made by a different decision-maker who has been given authority to act as proxy for the responsible decision authority.

Risk-informed decision making is distinguished from risk-based decision making in that RIDM is a fundamentally deliberative process that uses a diverse set of performance measures, along with other considerations, to inform decision making. The RIDM process acknowledges the role that human judgment plays in decisions, and that technical information cannot be the sole basis for decision making. This is not only because of inevitable gaps in the technical information, but also because decision making is an inherently subjective, values-based enterprise. In the face of complex decision making involving multiple competing objectives, the cumulative judgment provided by experienced personnel is an essential element for effectively integrating technical and nontechnical factors to produce sound decisions.

1.5.1 Part 1, Identification of Alternatives

In Part 1, *Identification of Alternatives,* objectives, which in general may be multifaceted and qualitative, are decomposed into their constituent-derived objectives, each of which reflects an individual issue that is significant to some or all of the stakeholders. At the lowest level of decomposition are ***performance objectives***, each of which is associated with a ***performance measure*** that quantifies the degree to which the performance objective is addressed by a given decision alternative. In general, a performance measure has a "direction of goodness" that indicates the direction of increasingly beneficial performance measure values. A comprehensive set of performance measures is considered collectively for decision making, reflecting stakeholder interests and spanning the mission execution domains of:

- Safety (e.g., avoidance of injury, fatality, or destruction of key assets)

- Technical (e.g., thrust or output, amount of observational data acquired)

- Cost (e.g., execution within allocated cost)

- Schedule (e.g., meeting milestones)

Objectives whose performance measure values must remain within defined limits for every feasible decision alternative give rise to ***imposed constraints*** that reflect those limits. Objectives and imposed constraints form the basis around which decision alternatives are compiled, and performance measures are the means by which their ability to meet imposed constraints and satisfy objectives is quantified.

1.5.2   Part 2, Risk Analysis of Alternatives

In Part 2, *Risk Analysis of Alternatives*, the performance measures of each alternative are quantified, taking into account any significant uncertainties that stand between the selection of an the alternative and the accomplishment of the objectives. Given the presence of uncertainty, the actual outcome of a particular decision alternative will be only one of a spectrum of forecasted outcomes, depending on the occurrence, nonoccurrence, or quality of occurrence of intervening events. Therefore, it is incumbent upon risk analysts to model each significant possible outcome, accounting for its probability of occurrence, in terms of the scenarios that produce it. This produces a distribution of outcomes for each alternative, as characterized by probability density functions (pdfs) over the performance measures (see Figure 6).

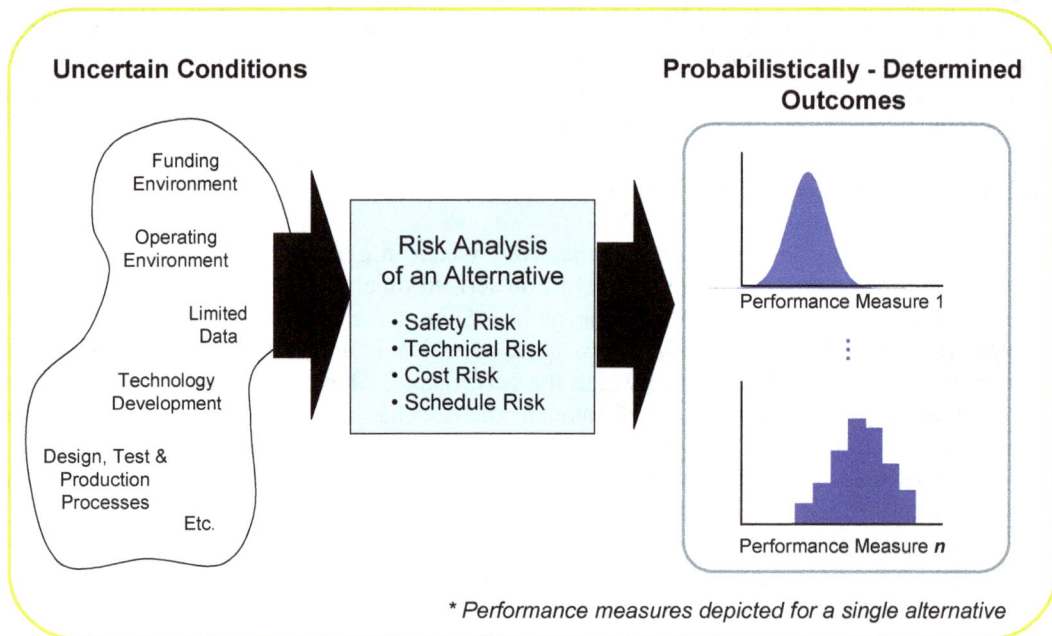

Figure 6. Uncertainty of Forecasted Outcomes Due to Uncertainty of Analyzed Conditions

RIDM is conducted using a graded approach, i.e., the depth of analysis needs to be commensurate with the stakes and complexity of the decision situations being addressed. Risk analysts conduct RIDM at a level sufficient to support robust selection of a preferred decision alternative. If the uncertainty on one or more performance measures is preventing the decision-maker from confidently assessing important differences between alternatives, then the risk analysis may be iterated in an effort to reduce uncertainty. The analysis stops when the technical case is made; if the level of uncertainty does not preclude a *robust decision* from being made then no further uncertainty reduction is warranted.

Performance Objectives, Performance Measures, and Imposed Constraints

In RIDM, top-level objectives, which may be multifaceted and qualitative, are decomposed into a set of ***performance objectives***, each of which is implied by the top-level objectives, and which cumulatively encompass all the facets of the top-level objectives. Unlike top-level objectives, each performance objective relates to a single facet of the top-level objectives, and is quantifiable. These two properties of performance objectives enable quantitative comparison of decision alternatives in terms of capabilities that are meaningful to the RIDM participants. Examples of possible performance objectives are:

- Maintain Astronaut Health and Safety   • Minimize Cost

- Maximize Payload Capability   • Maximize Public Support

A performance measure is a metric used to quantify the extent to which a performance objective is fulfilled. In RIDM, a performance measure is associated with each performance objective, and it is through performance measure quantification that the capabilities of the proposed decision alternatives are assessed. Examples of possible performance measures, corresponding to the above performance objectives, are:

- Probability of Loss of Crew (P(LOC))   • Cost ($)

- Payload Capability (kg)   • Public Support (1 – 5)

Note that, in each case, the performance measure is the means by which the associated performance objective is assessed. For example, the ability of a proposed decision alternative to Maintain Astronaut Health and Safety (performance objective) may be measured in terms of its ability to minimize P(LOC) (performance measure).

Although performance objectives relate to single facets of the top-level objectives, this does not necessarily mean that the corresponding performance measure is directly measurable. For example, P(LOC) might be used to quantify Maintain Astronaut Health and Safety, but the quantification itself might entail an assessment of vehicle reliability and abort effectiveness in the context of the defined mission profile.

An imposed constraint is a limit on the allowable values of the performance measure with which it is associated. Imposed constraints reflect performance requirements that are negotiated between NASA organizational units which define the task to be performed. In order for a proposed decision alternative to be feasible it must comply with the imposed constraints. A hard limit on the minimum payload capability that is acceptable is an example of a possible imposed constraint.

The principal product of the risk analysis is the ***Technical Basis for Deliberation (TBfD)***, a document that catalogues the set of candidate alternatives, summarizes the analysis methodologies used to quantify the performance measures, and presents the results. The TBfD is the input that risk-informs the deliberations that support decision making. The presence of this information does not necessarily mean that a decision is risk-informed; rather, without such information, a decision is not risk-informed. Appendix C contains a template that provides guidance on TBfD content. It is expected that the TBfD will evolve as the risk analysis iterates.

Robustness

A robust decision is one that is based on sufficient technical evidence and characterization of uncertainties to determine that the selected alternative best reflects decision-maker preferences and values given the state of knowledge at the time of the decision, and is considered insensitive to credible modeling perturbations and realistically foreseeable new information.

1.5.3 Part 3, Risk-Informed Alternative Selection

In Part 3, *Risk-Informed Alternative Selection*, deliberation takes place among the stakeholders and the decision-maker, and the decision-maker either culls the set of alternatives and asks for further scrutiny of the remaining alternatives OR selects an alternative for implementation OR asks for new alternatives.

To facilitate deliberation, a set of ***performance commitments*** is associated with each alternative. Performance commitments identify the performance that an alternative is capable of, at a given probability of exceedance, or risk tolerance. By establishing a risk tolerance for each performance measure independent of the alternative, comparisons of performance among the alternatives can be made on a risk-normalized basis. In this way, stakeholders and decision-makers can deliberate the performance differences between alternatives at common levels of risk, instead of having to choose between complex combinations of performance and risk.

It should be pointed out that while the ability to meet a specific performance commitment or requirement within a tolerable level of risk is a convenient means for comparing alternatives on a common basis, the decision maker also needs to be provided with information about other aspects of the performance measure uncertainty distributions such as the mean values and overall shapes of the pdfs. In particular, the full pdfs provide a means for examining the sensitivity of the ranking of alternatives to other choices of performance commitment thresholds and risk tolerance levels, which can be an important consideration in the decision making process. The ultimate decision among alternatives is best informed when a full spectrum of risk information is provided to the decision maker. This point will be discussed further in Section 3.3.2.6.

Deliberation and decision making might take place in a number of venues over a period of time or be tiered in a sequence of downselects. The rationale for the selected decision alternative is documented in a Risk-Informed Selection Report (RISR), in light of:

- The risk deemed acceptable for each performance measure

- The risk information contained in the TBfD

- The risk list

- The pros and cons of each contending decision alternative, as discussed during the deliberations.

Performance Commitments and Performance Requirements

A *performance commitment* is the performance measure value, at a given risk tolerance level for that performance measure, acceptable to the decision maker for the alternative that was selected. Performance commitments are used within the RIDM process in order to compare decision alternatives in terms of performance capability at specified risk tolerances for each performance measure (i.e., risk normalized).

Performance commitments serve as the starting point for the development of *performance requirements*, so that a linkage exists between the selected alternative, the risk tolerance of the decision-maker, and the requirements that define the objective to be accomplished. Performance commitments are not themselves performance requirements. Rather, performance commitments represent achievable levels of performance that are used to risk-inform the development of credible performance requirements as part of the overall systems engineering process. Risk tolerance is but one consideration that enters the requirements development process.

The figure below shows the evolution of a performance commitment in RIDM to a performance requirement in CRM. Performance Measure X is characterized by a pdf, due to uncertainties that affect the analyst's ability to forecast a precise value. In RIDM, the decision maker's risk tolerance is represented by the shaded "Risk" area on the left. In CRM, the assessed risk of not meeting the performance requirement is represented by the shaded "Risk" area on the right.

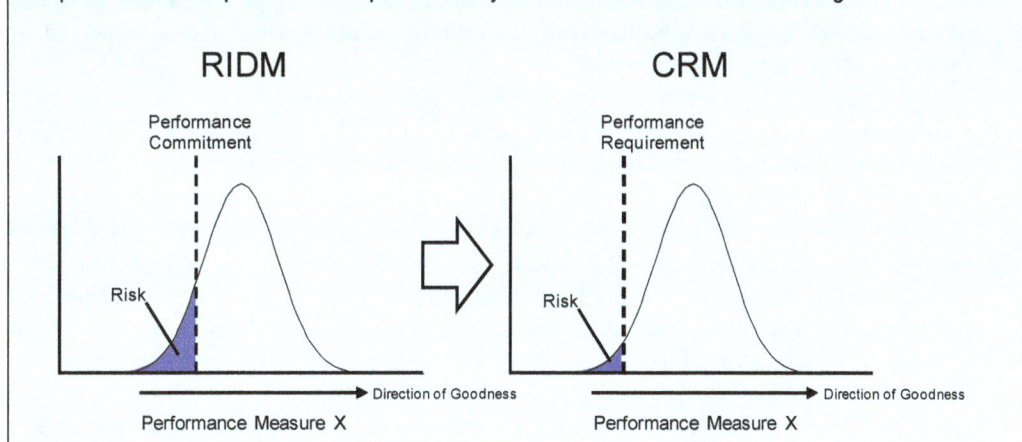

Guidance for the RISR is provided in Appendix D. This assures that deliberations involve discussion of appropriate risk-related issues, and that they are adequately addressed and integrated into the decision rationale.

1.5.4 Avoiding Decision Traps

Examination of actual decision processes shows a tendency for decision-makers to fall into certain *decision traps*. These traps have been categorized as follows [12]:

- **Anchoring** — This trap is the tendency of decision-makers to give disproportionate weight to the first information they receive, or even the first hint that they receive. It is related to a tendency for people to reason in terms of perturbations from a "baseline" perception, and to formulate their baseline quickly and sometimes baselessly.

- **Status Quo Bias** — There is a tendency to want to preserve the status quo in weighing decision alternatives. In many decision situations, there are good reasons (e.g., financial) to preserve the status quo, but the bias cited here is a more basic tendency of the way in which people think. Reference [12] notes that early designs of "horseless carriages" were strongly based on horse-drawn buggies, despite being sub-optimal for engine-powered vehicles. There is also the tendency for managers to believe that if things go wrong with a decision, they are more likely to be punished for having taken positive action than for having allowed the status quo to continue to operate.

- **Sunk-Cost** — This refers to the tendency to throw good money after bad: to try to recoup losses by continuing a course of action, even when the rational decision would be to walk away, based on the current state of knowledge. This bias is seen to operate in the perpetuation of projects that are floundering by any objective standard, to the point where additional investment diverts resources that would be better spent elsewhere. A decision process should, in general, be based on the current situation: what gain is expected from the expenditure being contemplated.

- **Confirmation Bias** — This refers to the tendency to give greater weight to evidence that confirms our prior views, and even to seek out such evidence preferentially.

- **Framing** — This refers to a class of biases that relate to the human tendency to respond to how a question is framed, regardless of the objective content of the question. People tend to be risk-averse when offered the possibility of a sure gain, and risk-seeking when presented with a sure loss. However, it is sometimes possible to describe a given situation either way, which can lead to very different assessments and subsequent decisions.

- **Overconfidence** — This refers to the widespread tendency to underestimate the uncertainty that is inherent in the current state of knowledge. While most "experts" will acknowledge the presence of uncertainty in their assessments, they tend to do a poor job of estimating confidence intervals, in that the truth lies outside their assessed bounds much more often than would be implied by their stated confidence in those bounds. This is particularly true for decisions that are challenging to implement, as many decisions at NASA are. In the face of multiple sources of uncertainty, people tend to pay attention to the few with which they have the most experience, and neglect others. It is also particularly true for highly unlikely events, where there are limited data available to inform expert judgment.

- **Recallability** — This refers to the tendency of people to be strongly influenced by experiences or events that are easier for them to recall, even if a neutral statistical analysis of experience would yield a different answer. This means that dramatic or extreme events may play an unwarrantedly large role in decision making based on experience.

The RIDM process helps to avoid such traps by establishing a rational basis for decision-making, ensuring that the implications of each decision alternative have been adequately analyzed, and by providing a structured environment for deliberation in which each deliberator can express the merits and drawbacks of each alternative in light of the risk analysis results.

1.6   Overview of the CRM Process

Once an alternative has been selected using the RIDM process and performance requirements have been developed for it as part of the technical requirements definition process described in the Systems Engineering Handbook [2], the risk associated with its implementation is managed using the CRM process. Because CRM takes place in the context of explicitly-stated performance requirements, the risk that the CRM process manages is the potential for performance shortfalls, which may be realized in the future, with respect to these requirements.

The CRM process described herein starts from the five cyclical functions of *Identify*, *Analyze*, *Plan*, *Track*, and *Control*, supported by comprehensive *Communicate* and *Document* functions [13]. This process is shown graphically in Figure 7.

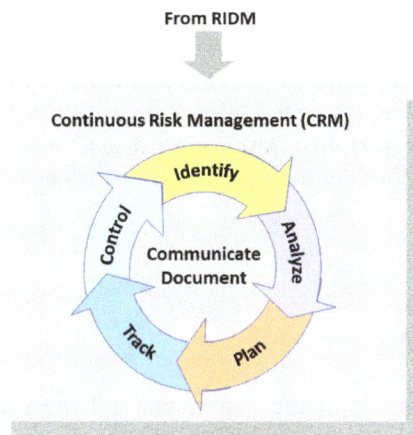

Figure 7. The CRM Process

The CRM process manages risk by identifying specific issues that are of concern to one or more stakeholders, and which are perceived as presenting a risk to the achievement of one or more performance requirements. These issues are referred to as ***individual risks***, and collectively represent the set of undesirable scenarios that put the achievement of the activity's performance requirements at risk. Thus, each performance requirement has an associated ***performance risk*** that results from the aggregation of individual risks. The aggregation usually involves more than just a simple addition of the individual risks since it must account for any interactions and dependencies that exist among them.

Performance risk is quantified using a risk model, most likely developed starting from that developed for the risk analysis of the selected alternative conducted during the RIDM process but expanded as needed to address all significant risk issues. The scenarios in the RIDM risk model that lead to performance shortfalls are initially captured as individual risks to be managed by the CRM process. As implementation proceeds and additional individual risks are identified, additional scenarios may be added to the model in order to analyze their effect on each performance requirement's performance risk as well. This allows the management of individual risks to be prioritized in terms of their significance with respect to performance risk.

The CRM process is portrayed in this handbook primarily as an iterated sequence of steps. This sequence generally holds true for a given individual risk, but different individual risks can be at different steps in the process, so that at any given time, any or all of the steps may be in effect for one or more individual risks. In addition, while the process is more-or-less sequential as shown in Figure 7, the steps can be approached non-sequentially when needed. This could include situations, for example, where the individual risks need to be analyzed at a high level to support near-term planning before being analyzed at a more detailed level to support longer-term planning. More on this subject will be presented in Section 4.3.

1.6.1   Step 1, Identify

The purpose of the *Identify* step is to capture stakeholders' concerns regarding the achievement of performance requirements. These concerns are captured as individual risks in a **risk database**. Each individual risk is articulated as a **risk statement** that contains a *condition*, a *departure*, an *asset*, and a *consequence*. Each of these elements is defined in the blue box below.

Individual risks also include a **narrative description** that allows the identifier of the individual risk to provide background information considered relevant to understanding and appreciating the identifier's concerns.

1.6.2   Step 2, Analyze

The objectives of the *Analyze* step are:

- To estimate the likelihoods of the departures and the magnitudes of the consequence components of individual risks, including timeframe, uncertainty characterization, and quantification

- To assign, in a timely fashion, a criticality rank to each individual risk based on:

    o   The probability that the departure will occur

o   The magnitude of the consequence given occurrence of the departure[3]

o   The point in the activity's timeline when the individual risk first surfaced (e.g., PDR, CDR)

o   The magnitude of the uncertainties

o   The amount of time available after the condition is identified before a departure can possibly occur.

- To update the performance risk to incorporate new individual risks or changes in existing individual risks

- To determine which departure events and parameters within the models are the most important contributors to each performance risk.

Individual Risks

An individual risk is a specific issue of concern to one or more stakeholders, which is perceived as presenting a risk to the achievement of one or more performance requirements. Collectively, individual risks represent the identified set of undesirable scenarios that put the achievement of performance requirements at risk. Individual risks are captured in a risk database in terms of the following elements:

- *Risk Statement* - A concise description of an individual risk that can be understood and acted upon. Risk statements have the following structure: "Given that [CONDITION], there is a possibility of [DEPARTURE] adversely impacting [ASSET], which can result in [CONSEQUENCE]."

    - The CONDITION is a single phrase that describes the current key fact-based situation or environment that is causing concern, doubt, anxiety, or uneasiness.

    - The DEPARTURE describes a possible change from the (agency, program, project, or activity) baseline project plan. It is an undesired event that is made credible or more likely as a result of the CONDITION.

    - The ASSET is an element of the organizational unit portfolio (OUP) (analogous to a WBS). It represents the primary resource that is affected by the individual risk.

    - The CONSEQUENCE is a single phrase that describes the foreseeable, credible negative impact(s) on the organizational unit's ability to meet its performance requirements.

- *Narrative Description* – Additional detail regarding the events, circumstances, and interrelationships within the activity that may affect the individual risk. This description is more detailed than can be captured in the risk statement.

---

[3] In the context of CRM, consequence is defined as the failure to meet a performance requirement. Possible measures of consequence that can be used in CRM will be discussed in Section 4.2.1.1.

The *Analyze* step assures that the issues underlying each individual risk are properly understood and evaluated in terms of its impact on performance risk, as illustrated in Figure 8. *Analyze* enables prioritization of individual risks so they can be responded to in a manner that efficiently and effectively supports program/project success.

Performance Risk

Performance risk is the probability of not meeting a performance requirement. Each performance requirement has an associated performance risk that is produced by those individual risks which, in the aggregate, threaten the achievement of the requirement. Quantification of a performance requirement's performance risk is accomplished by means of a scenario-based risk model that incorporates the individual risks so that their aggregate effect on the forecasted probabilities of achieving the performance requirements can be analyzed.

The term *performance risk* is essentially synonymous to the term *risk*. Both refer to the potential for shortfalls with respect to performance requirements. The term *performance risk* is used during CRM to distinguish it from the individual risks, and to emphasize its association with the performance requirements.

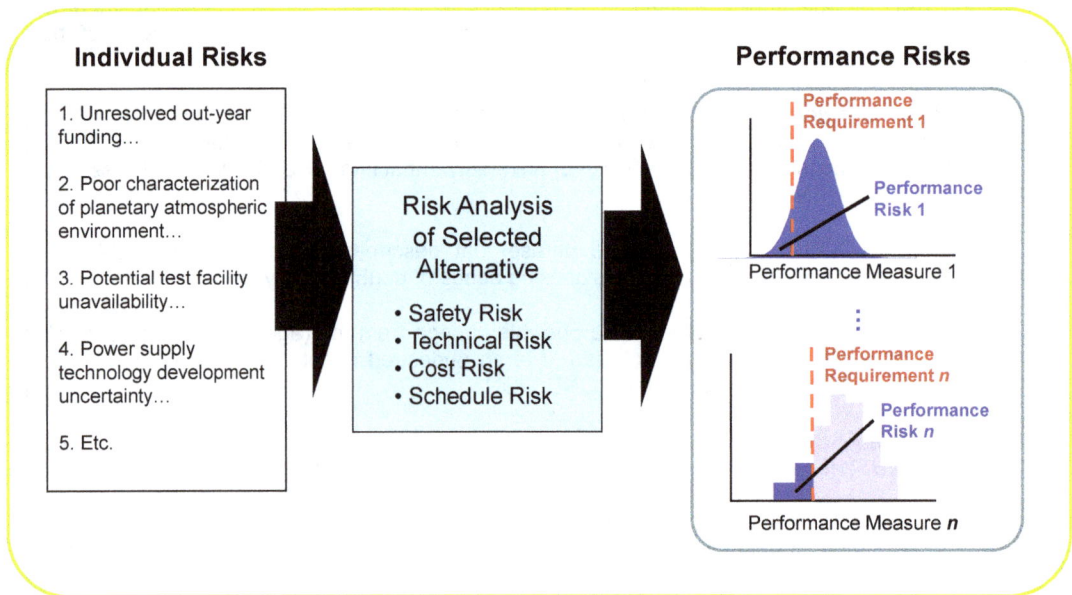

Figure 8. Integration of Individual Risks to Produce Performance Risks

1.6.3 Step 3, Plan

The objective of the *Plan* step is to decide what action, if any, should be taken to reduce the performance risks that are caused by the aggregation of identified individual risks. The possible actions are:

- *Accept* – A certain level of performance risk can be accepted if it is within the risk tolerance of the program/project manager.

- *Mitigate* – Mitigation actions can be developed which address the drivers of the performance risk.

- *Watch* – Risk drivers can be selected for detailed observation and contingency plans developed.

- *Research* – Research can be conducted to better understand risk drivers and reduce their uncertainties.

- *Elevate* – Risk management decisions should be elevated to the sponsoring organization at the next higher level of the NASA hierarchy when performance risk can no longer be effectively managed within the present organizational unit.

- *Close* – An individual risk can be closed when all associated risk drivers are no longer considered potentially significant.

Selection of an appropriate risk management action is supported by risk analysis of alternatives and subsequent deliberation, using the same general principles of risk-informed decision making that form the basis for the RIDM process.

1.6.4 Step 4, Track

The objective of the *Track* step is to acquire, compile, and report observable data to track the progress of the implementation of risk management decisions, and their effectiveness once implemented. The tracking task of CRM serves as a clearing house for new information that could lead to any of the following:

- A new risk item

- A change in risk analysis

- A change in a previously agreed-to plan

- The need to implement a previously agreed-to contingency.

1.6.5 Step 5, Control

When tracking data indicates that a risk management decision is not impacting risk as expected, it may be necessary to implement a *control* action. Control actions are intended to assure that the planned action is effective. If the planned action becomes unviable, due either to an inability to implement it or a lack of effectiveness, then the Plan step is revisited and a different action is chosen.

1.6.6 Communicate and Document

Communication and documentation are key elements of a sound and effective CRM process. Well-defined, documented communication tools, formats, and protocols assure that:

- Individual risks are identified in a manner that supports the evaluation of their impacts on performance risk

- Individual risks that impact multiple organizational units (i.e., crosscutting risks) are identified, enabling the coordination of risk management efforts

- Performance risks, and associated risk drivers, are reported by each organizational unit to the sponsoring organization at the next higher level of the NASA hierarchy in a manner that allows the higher level organization to integrate that information into its own assessment of performance risk relative to its own performance requirements

- Risk management decisions and their rationales are captured as part of the institutional knowledge of the organization.

2   RIDM PROCESS INTERFACES

As discussed in Section 1, within each NASA organizational unit, RIDM and CRM are integrated into a coherent RM framework in order to:

- Foster proactive risk management

- Better inform decision making through better use of risk information

- More effectively manage implementation risks by focusing the CRM process on the performance requirements emerging from the RIDM process.

The result is a RIDM process within each unit that interfaces with the unit(s) at the next higher and lower levels in the organizational hierarchy when negotiating objectives and establishing performance requirements, as well as with its own unit's CRM process during implementation. This situation is illustrated graphically in Figure 9, which has been reproduced from NPR 8000.4A.[4] The following subsections discuss these interfaces in more detail.

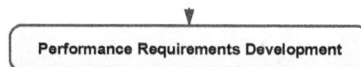

Figure 9. Coordination of RIDM and CRM within the NASA Hierarchy (Illustrative)

---

[4] Figure 5 in NPR 8000.4A.

2.1 Negotiating Objectives across Organizational Unit Boundaries

Organizational units negotiate with the unit(s) at the next lower level, including center support units, a set of objectives, deliverables, performance requirements, performance measures, resources, and schedules that defines the tasks to be performed. These elements reflect the outcome of the RIDM process that has been conducted by the level above and the execution of its own responsibility to meet the objectives to which it is working.

- The organizational unit at the level above is responsible for assuring that the objectives and imposed constraints assigned to the organizational unit at the lower level reflect appropriate tradeoffs between and among competing objectives and risks. Operationally, this means that a linkage is maintained to the performance objectives used in the RIDM process of the unit at the higher level. It also means that the rationale for the selected alternative is preserved, in terms of the imposed constraints that are accepted by the unit at the lower level.

- The organizational unit at the level below is responsible for establishing the feasibility and capability of accomplishing the objectives within the imposed constraints, and managing the risks of the job it is accepting (including identification of mission support requirements).

Additional discussion related to objectives can be found in Section 3.1.1.

2.2 Preparing a Preliminary Risk Management Plan

A preliminary risk management plan (RMP) is prepared before RIDM is carried out to help guide the RIDM process. When RIDM at a project level is initiated, the RMP at the next higher level (i.e., program level) should already have been written. The requirements contained within the RMP at program level will serve as the basis for conducting RIDM at the project level. The preliminary RMP for the project is prepared in order to document how performance requirements at the program level will be implemented at the project level.

The project level RMP cannot be completely developed until RIDM for the project has been completed, the design alternative to be pursued has been selected, and the detailed performance requirements corresponding to the selected design alternative have been developed by systems engineering. Therefore, a discussion of the contents of the final RMP is deferred until Section 4.1.1, as part of the CRM Initialization topic. The reader may refer ahead to Section 4.1.1 for this information, since the contents of the preliminary RMP will be similar to the contents of the final RMP (only the level of detail will be less).

2.3 Coordination of RIDM and CRM

RIDM and CRM are complementary RM processes that operate within every organizational unit. Each unit applies the RIDM process to decide how to meet objectives and applies the CRM process to manage risks associated with implementation.[5] In this way, RIDM and CRM work together to provide comprehensive risk management throughout the entire life cycle of the project. The following subsections provide an overview of the coordination of RIDM and CRM. Additional information can be found in Section 4.

2.3.1 Initializing the CRM Risks Using the Risk Analysis of the Selected Alternative

For the selected alternative, the risk analysis that was conducted during RIDM represents an initial identification and assessment of the scenarios that could lead to performance shortfalls. These scenarios form the basis for an initial risk list that is compiled during RIDM for consideration by the decision-maker. Upon implementation of the selected alternative, this information is available to the CRM process to initialize its *Identify*, *Analyze,* and *Plan* activities. Figure 10 illustrates the situation. The scenarios of the risk model are input to the *Identify* activity. The effects that these scenarios have on the ability to meet the performance requirements are assessed in the *Analyze* activity. This activity integrates the scenario-based risk analysis from RIDM into the CRM analysis activities as a whole, in the context of the performance requirements to which CRM is managing.

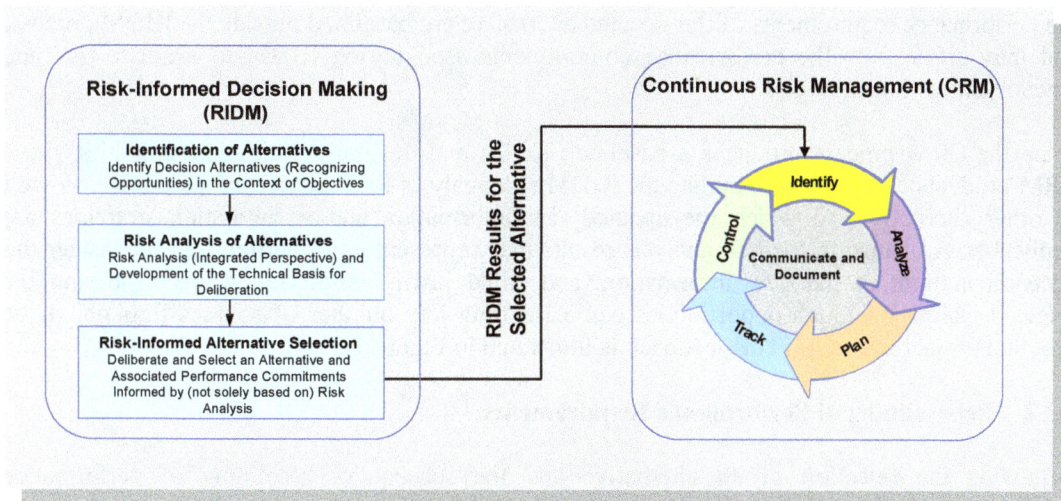

Figure 10. RIDM Input to CRM Initialization

During the initialization process for CRM, and subsequently during the project timeline, there will be a shift in emphasis from parameter based risk models to scenario based risk models. During RIDM, many of the risk models will tend to be parameter based. This is particularly true of cost and mass growth models, where the total cost of a task will be accumulated from the individual costs of the constituent subtasks and the total mass of a subsystem will be

---

[5] In the context of CRM, the term "risk" is used to refer to a family of scenarios potentiated by a particular identifiable underlying condition that warrants risk management attention, because it can lead to performance shortfalls. This usage is more specific than the operational definition of risk presented in Section 1.3, and is formulated so that the underlying conditions can be addressed during implementation.

accumulated from the masses of the individual components. Uncertainty distributions for these constituent costs and masses, in turn, will typically be determined from historical experience that tends to have a generic flavor. By contrast, scenario based models, such as event trees and fault trees, account more directly for project-specific conditions and departures that lead directly to performance risks. The transition to scenario-based models is necessitated by the objective of CRM, which is to produce response options such as mitigation or research that resolve project-specific risks.

Strategies for addressing risks and removing threats to requirements are developed in the *Plan* activity, and are also informed by the RIDM risk analysis. While the RIDM risk analysis of the selected alternative *informs* CRM, it does not replace the need for independent CRM *Identify*, *Analyze,* and *Plan* activities. There are many reasons for this, but one key reason is that the risk analysis in RIDM is conducted expressly for the purposes of distinguishing between alternatives and generating performance commitments, not for the purpose of managing risk during implementation. Therefore, for example, uncertainties that are common to all alternatives and that do not significantly challenge imposed constraints will typically not be modeled to a high level of detail since they do not serve to discriminate between alternatives or affect the feasibility of the alternative. They will instead be modeled in a more simple and conservative manner. Also, the performance requirements of the selected alternative are baselined outside the RIDM process, and may differ from the performance commitments used during RIDM to evaluate risk and develop mitigation strategies.

Once the CRM process produces a baseline risk list and develops mitigation strategies, these CRM products can be used to update the RIDM risk analysis for the selected alternative, as well as other alternatives to which the updated risk information and/or mitigation strategies are applicable. A change in the risk analysis results may represent an opportunity to reconsider the decision in light of the new information, and could justify modifying or reconsidering the selected alternative. Such opportunities can arise from any number of sources throughout the program/project life cycle. This feedback is illustrated in Figure 11.

2.3.2 Rebaselining of Performance Requirements

Following the selection of an alternative and the subsequent baselining of performance requirements, CRM operates at each level of the NASA hierarchy to implement the selected alternative in compliance with the performance requirements at that level. Ideally, CRM will operate smoothly to achieve the objectives without incident. However, there are two general classes of circumstance which entail rebaselining of the requirements:

- Issues can arise that make managing the risk of the selected alternative untenable, and rebaselining of the requirements that the organizational unit is working to becomes necessary. This might be due to:

  o A newly identified risk-significant scenario for which no mitigation is available within the scope of the current requirements

  o An emerging inability to control a previously identified risk.

When this occurs, the decision for managing the issue is elevated as appropriate within the CRM process. The unit at the higher level then has options at its disposal to support the efforts of the unit at the lower level, such as releasing management margin in some resource (e.g., funding, mass, schedule), relaxing one or more performance requirements, or re-executing the RIDM process to develop a new set of performance requirements that are achievable given the current situation. Each of these options represents various degrees of rebaselining.

- Issues can arise in which an organizational unit is able to manage its risk, but can do so only by modifying the selected alternative to a degree that results in a modification of the derived requirements flowing down to the unit(s) below, requiring a rebaselining of the derived requirements imposed on those units.

In each of the above situations, when the degree of modification to the selected alternative is sufficiently large, the RIDM process should be invoked to produce an updated alternative to serve as the basis for rebaselined requirements. The rebaselining may involve an adjustment process, wherein certain requirements are modified to make them more applicable and practicable, or alternatively an outright waiving of requirements that are unnecessary or counterproductive. This situation is shown in Figure 11. As indicated by the figure, the RIDM process is entered at Part 1, Identification of Alternatives, which addresses the development of performance measures and imposed constraints, as well as the compilation of a set of alternatives for analysis. In general, it is not expected that the performance measures will change, so RIDM is executed using those derived from the existing objectives hierarchy. However, there may be cause to modify an imposed constraint, particularly if it relates to the threatened requirement(s) and if modification/relaxation produces feasible requirements at a tolerable impact to objectives.

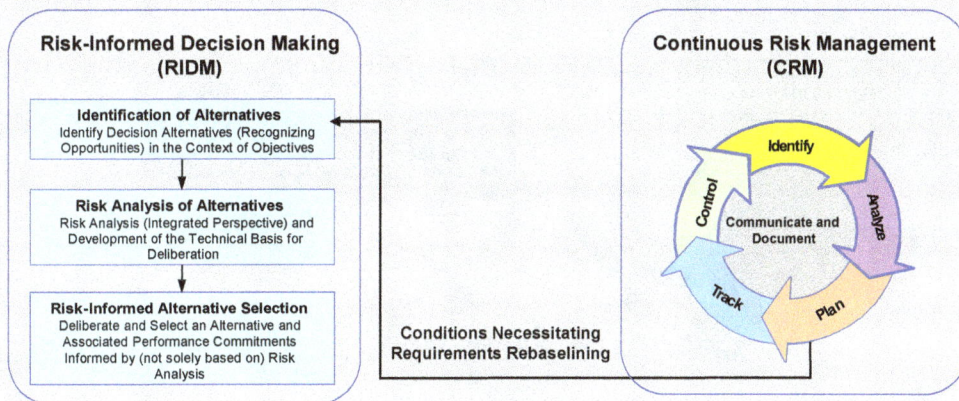

Figure 11. Rebaselining of Performance Requirements

The set of decision alternatives compiled for risk analysis will typically differ from the set analyzed initially, primarily in its scope. It is expected that only in rare circumstances will the entirety of the selected alternative be subject to re-evaluation. Instead, the scope of RIDM should be as narrow as practical while also not reflecting a *sunk cost* or *status quo* mindset (discussed in Section 1.5.4). Within this scope, alternatives that were previously shown to be unattractive can be excluded if they are unaffected by the circumstances surrounding the rebaselining, but circumstances might also suggest alternatives that weren't considered before; care should be taken to identify these alternatives, and not draw only from the previous set.

Rebaselining is done in light of current conditions. These conditions include not only the circumstances driving the rebaselining, but also those of the activity in general, such as budget status and accomplishments to date. Once the new set of decision alternatives is identified, RIDM proceeds as usual, taking advantage of the previous risk analysis to the extent practical given the new set and the current program/project status.

The effects of requirements rebaselining are not confined to the organization that experienced the original CRM risk. Every organization in the NASA hierarchy whose requirements are derived from the rebaselined requirements is potentially affected. The scope of affected organizations depends on the level at which the risk originates, the number of levels that the risk management decision is elevated before it can be mitigated within the existing requirements of the unit to which it is elevated, and the particulars of any changes to the mitigating unit's flowed-down requirements. Figure 12 illustrates the situation of rebaselining following elevation, showing the potential scope of the rebaselined requirements as they flow down to subordinate organizational units.

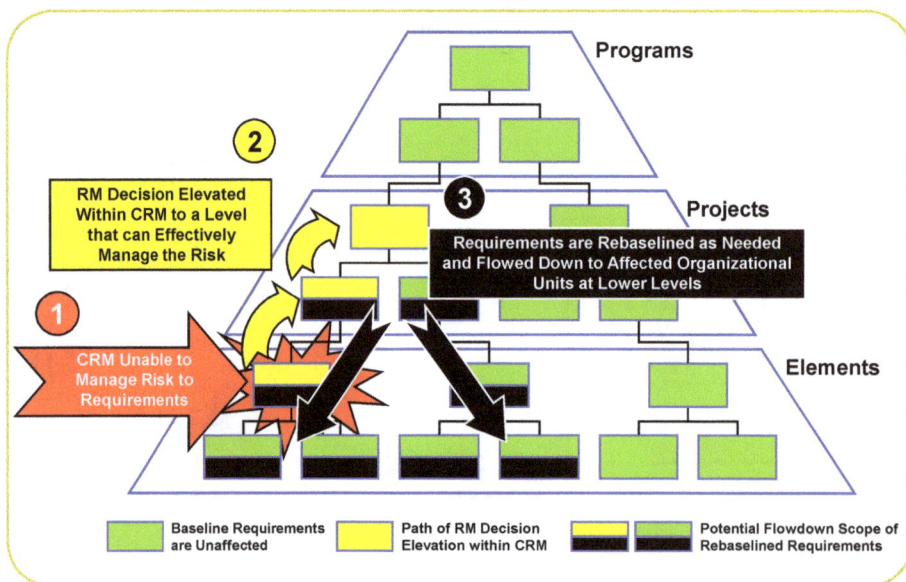

Figure 12. Scope of Potentially Affected Organizations Given Rebaselining

In certain instances, new information may emerge that represents an opportunity to rethink a previous decision. Just such a situation was mentioned in the case where the CRM process produces a mitigation strategy that, if retroactively applied to the set of candidate decision alternatives, could shift the preferred alternative from the selected alternative to a different one. Other opportunities can arise from ancillary analyses conducted either internally or externally to NASA, technology advancements, test results, etc.

2.4   Maintaining the RIDM Process

The discussion of RIDM interfaces in the previous sections shows the importance of maintaining a functioning RIDM capability throughout the program/project life cycle. This capability includes:

- Reviewable TBfD and RISR documents containing the rationale for prior decision making and the discussion of issues of significance to the stakeholders.

- Accessible *objectives hierarchies* (discussed in Section 3.1.1) to serve as the sources of relevant performance measures or as the anchor points for decomposing objectives to finer levels of resolution. This assures that decisions remain tied to NASA strategic goals.

- Accessible risk analysis framework structures and risk models that were used to quantify the performance measures.

- The ability, at every organizational level, to integrate information from lower levels to support RIDM processes that reflect current conditions throughout the NASA hierarchy. This includes program/project status details as well as relevant analyses.

- Access to relevant discipline-specific analyses to use as input to risk analysis, as well as access to relevant expertise to support additional discipline-specific analyses needed for decision making.

- Maintenance of risk analysis expertise to coordinate the development of risk information and integrate it into the TBfD.

3   THE RIDM PROCESS

Figure 13 expands the three parts of RIDM into a sequence of six process steps.

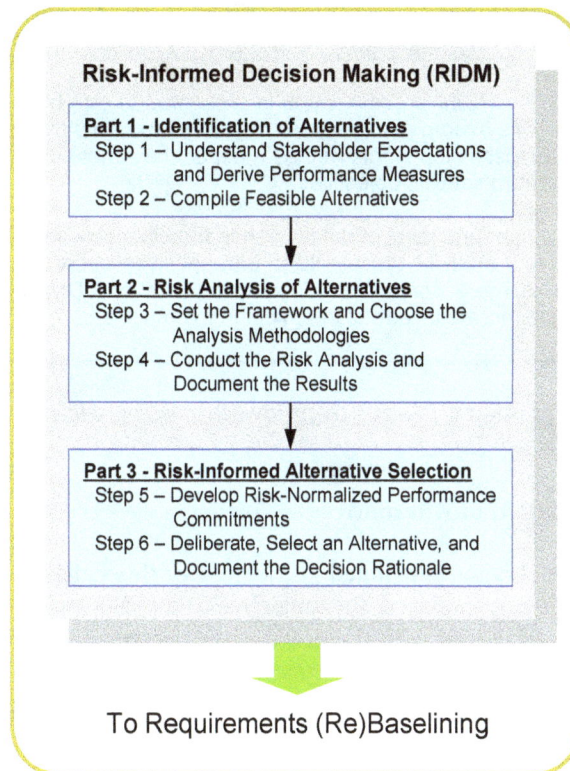

Risk-Informed Decision Making (RIDM)

Part 1 - Identification of Alternatives
Step 1 – Understand Stakeholder Expectations and Derive Performance Measures
Step 2 – Compile Feasible Alternatives

Part 2 - Risk Analysis of Alternatives
Step 3 – Set the Framework and Choose the Analysis Methodologies
Step 4 – Conduct the Risk Analysis and Document the Results

Part 3 - Risk-Informed Alternative Selection
Step 5 – Develop Risk-Normalized Performance Commitments
Step 6 – Deliberate, Select an Alternative, and Document the Decision Rationale

To Requirements (Re)Baselining

Figure 13. RIDM Process Steps

It is important to note that although Figure 4 and Figure 13 depict the RIDM process as a linear sequence of steps, in practice it is expected that some steps could overlap in time and that the process is iterative. Information from latter steps feeds back into progressively more refined execution of previous steps until stakeholder issues are adequately addressed and the decision-maker has sufficient information, at a sufficient level of analytical rigor, to make a robust risk-informed decision. The primary issues driving the need for iteration are discussed in the following subsections, in the context of the RIDM process steps in which they arise.

The RIDM process has been informed by current theoretical and practical work in decision analysis and analytic-deliberative processes (see, for example, [14], [15], [16]). Some methodological tools and techniques, such as objectives hierarchies, performance measures, and deliberation, have been adopted into the RIDM process as being generally applicable to structured, rational decision making. Others, such as analytic hierarchy process (AHP) and multi-attribute utility theory (MAUT), are formally applicable to rational decision making but also present practical challenges in the context of requirements development within a complex organizational hierarchy having its own highly developed program management policies and practices. It is left to the discretion of the practitioner to determine, on a case-by-case basis, whether or not such techniques will aid in deliberation and selection of a decision alternative.

Planetary Science Mission Example

An example application of the RIDM process steps is presented in this handbook for a hypothetical planetary science mission. This example is contained in yellow boxes distributed throughout Section 3. Discussion of each RIDM process step is followed by a notional example of how it might be applied to a specific decision, derived from specific objectives.

The methods, measures, scope, and level of detail used in the planetary science mission example are not meant to prescribe how the RIDM process is to be applied in every instance. Rather, they are meant to give the reader a more concrete understanding of the RIDM process and its practical application, in addition to the more conceptual treatment in the main text.

The sections that follow provide a process overview, discussing each of the main activities that support each step.

3.1   Part 1 – Identification of Alternatives

As indicated in NPR 8000.4A and illustrated in Figure 4 of this handbook, decision alternatives are identifiable only in the context of the objectives they are meant to satisfy. Therefore, identification of alternatives begins with the process of understanding stakeholder expectations. From there, a basis for evaluating decision alternatives is developed by decomposing stakeholder expectations into quantifiable objectives that enable comparison among the candidates. Only then, after an appropriate context has been established, is it possible to compile a set of feasible alternatives that address the objectives. Figure 14 illustrates this part of the process, which is delineated in subsequent subsections.

3.1.1   Step 1 – Understand Stakeholder Expectations and Derive Performance Measures

3.1.1.1 Understand Stakeholder Expectations

The development of unambiguous performance measures and imposed constraints, reflecting stakeholder expectations, is the foundation of sound decision making. Paragraph 3.2.1 of NPR 7123.1A establishes systems engineering process requirements for stakeholder expectations definition, and the NASA Systems Engineering Handbook provides further guidance on understanding stakeholder expectations.

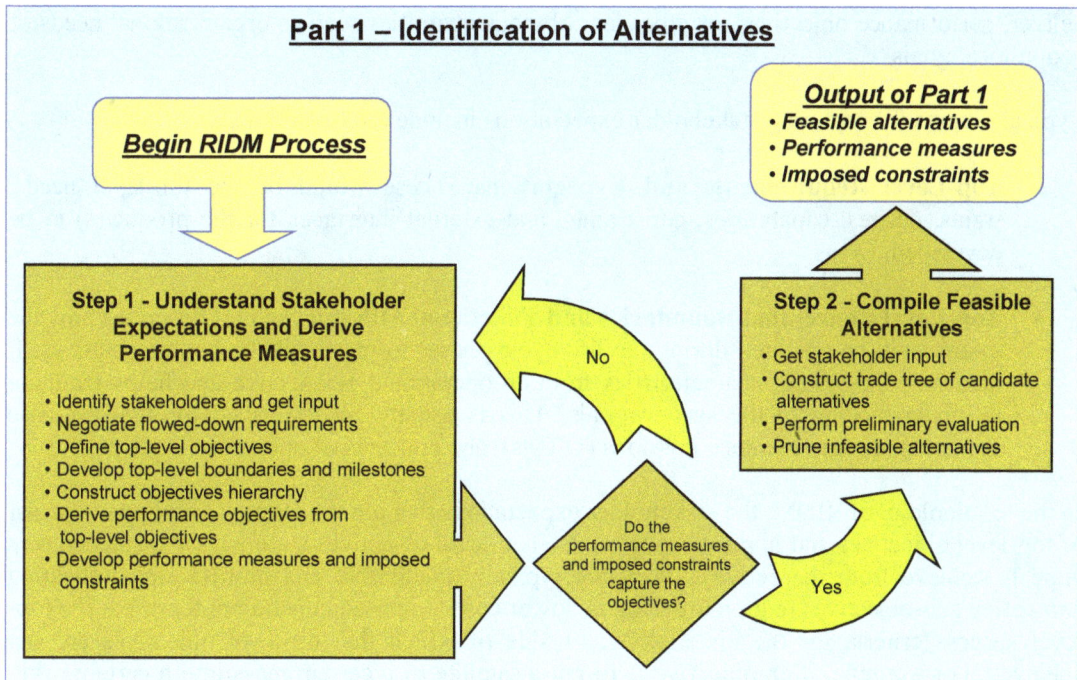

Figure 14. RIDM Process Flowchart: Part 1, Identification of Alternatives

Typical inputs needed for the stakeholder expectations definition process include:

- **Upper Level Requirements and Expectations:** These would be the requirements and expectations (e.g., needs, wants, desires, capabilities, constraints, external interfaces) that are being flowed down to a particular system of interest from a higher level (e.g., program, project, etc.).

- **Stakeholders:** Individuals or organizations that are materially affected by the outcome of a decision or deliverable.[6]

A variety of organizations, both internal and external to NASA, may have a stake in a particular decision. Internal stakeholders might include NASA Headquarters (HQ), the NASA Centers, and NASA advisory committees. External stakeholders might include the White House, Congress, the National Academy of Sciences, the National Space Council, and many other groups in the science and space communities.

Stakeholder expectations, the vision of a particular stakeholder individual or group, result when they specify what is desired as an end state or as an item to be produced and put bounds upon the achievement of the goals. These bounds may encompass expenditures (resources), time to

---

[6] The National Research Council (NRC) defines stakeholders as being persons who are external to the organization doing the work or making the decision [17], This handbook, however, uses the term to include both internal and external entities who have a stake in the decision.

deliver, performance objectives, or other less obvious quantities such as organizational needs or geopolitical goals.

Typical outputs for capturing stakeholder expectations include the following:

- **Top-Level Requirements and Expectations:** These would be the top-level needs, wants, desires, capabilities, constraints, and external interfaces for the product(s) to be developed.

- **Top-Level Conceptual Boundaries and Functional Milestones:** This describes how the system will be operated during the life cycle phases to meet stakeholder expectations. It describes the system characteristics from an operational perspective and helps facilitate an understanding of the system goals. This is usually accomplished through use-case scenarios, design reference missions (DRMs), and concepts of operation (ConOps).

In the terminology of RIDM, the stakeholder expectations that are the outputs of this step consist of top-level objectives and imposed constraints. Top-level objectives state what the stakeholders hope to achieve from the activity. They are typically qualitative and multifaceted, reflecting competing sub-objectives (e.g., more data vs. lower cost). Imposed constraints represent the top-level success criteria for the undertaking, outside of which the top-level objectives are not achieved. For example, if an objective is to put a satellite of a certain mass into a certain orbit, then the ability to loft that mass into that orbit is an imposed constraint, and any proposed solution that is incapable of doing so is infeasible.

3.1.1.2 Derive Performance Measures

In general, decision alternatives cannot be directly assessed relative to multifaceted and/or qualitative top-level objectives. Although the top-level objectives state the goal to be accomplished, they may be too complex, as well as vague, for any operational purpose. To deal with this situation, objectives are decomposed, using an *objectives hierarchy*, into a set of conceptually distinct lower-level objectives that describe the full spectrum of necessary and/or desirable characteristics that any feasible and attractive alternative should have. When these objectives are quantifiable via performance measures, they provide a basis for comparing proposed alternatives.

Constructing an Objectives Hierarchy

An objectives hierarchy is constructed by subdividing an objective into lower-level objectives of more detail, thus clarifying the intended meaning of the general objective. Decomposing an objective into precise lower-level objectives clarifies the tasks that must be collectively achieved and provides a well-defined basis for distinguishing between alternative means of achieving them.

Planetary Science Mission Example: Understand Stakeholder Expectations

The Planet "X" Program Office established an objective of placing a scientific platform in orbit around Planet "X" in order to gather data and transmit it back to Earth. Stakeholders include:

- The planetary science community who will use the data to further humanity's understanding of the formation of the solar system

- The Earth science community who will use the data to refine models of terrestrial climate change and geological evolution

- Environmental groups who are concerned about possible radiological contamination of Planet "X" in the event of an orbital insertion mishap

- Mission support offices (MSOs) who are interested in maintaining their infrastructure and workforce capabilities in their areas of specialized expertise

Specific expectations include:

- The envisioned ConOps is for a single launch of a scientific platform that will be placed in a polar orbit around Planet X

- The envisioned scientific platform will include a radioisotope thermoelectric generator (RTG) for electrical power generation

- The launch date must be within the next 55 months due to the launch window

- The scientific platform should provide at least 6 months of data collection

- Data collection beyond the initial 6 months is desirable but not mission critical

- The scientific platform will include a core data collection capability in terms of data type and data quality (for the purpose of this example, the specifics of the data are unspecified)

- Collection of additional (unspecified) types of scientific data is desirable if the capability can be provided without undue additional costs or mission success impacts

- The mission should be as inexpensive as possible, with a cost cap of $500M

- The probability of radiological contamination of Planet "X" should be minimized, with a goal of no greater than 1 in 1000 (0.1%)

An objectives hierarchy is shown notionally in Figure 15. At the first level of decomposition the top-level objective is partitioned into the NPR 8000.4A mission execution domains of Safety, Technical, Cost, and Schedule. This enables each performance measure and, ultimately, performance requirement, to be identified as relating to a single domain. Below each of these domains the objectives are further decomposed into sub-objectives, which themselves are iteratively decomposed until appropriate quantifiable performance objectives are generated.

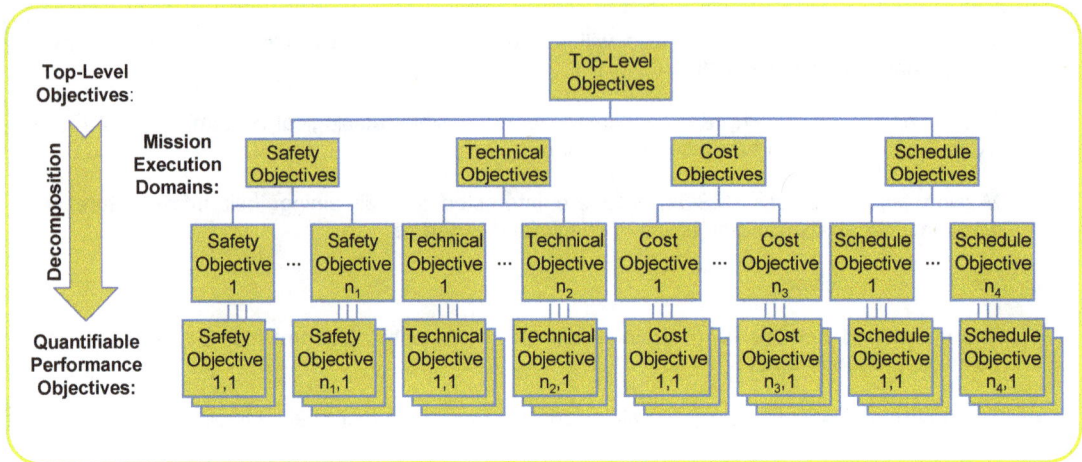

Figure 15. Notional Objectives Hierarchy

There is no prescribed depth to an objectives hierarchy, nor must all performance objectives reside at the same depth in the tree. The characteristics of an objectives hierarchy depend on the top-level objective and the context in which it is to be pursued. Furthermore, a unique objectives hierarchy is not implied by the specification of an objective; many different equally legitimate objectives hierarchies could be developed.

When developing an objectives hierarchy there is no obvious stopping point for the decomposition of objectives. Judgment must be used to decide where to stop by considering the advantages and disadvantages of further decomposition. Things to consider include:

- Are all facets of each objective accounted for?

- Are all the performance objectives at the levels of the hierarchy quantifiable?[7]

- Is the number of performance objectives manageable within the scope of the decision-making activity?

---

[7] The question about all the performance objectives being quantifiable does not preclude the fact that some are likely to be fundamentally qualitative in nature. For example, concerns of a commercial or international partner may pertain to the flexibility of a system or the missions to be conducted using it. One way to address qualitative objectives in quantitative terms is to use proxy scales and/or constructed scales. The application of proxy and constructed scales for this purpose is discussed later in this section.

One possibility is to use a "test of importance" to deal with the issue of how broadly and deeply to develop an objectives hierarchy and when to stop. Before an objective is included in the hierarchy, the decision-maker is asked whether he or she feels the best course of action could be altered if that objective were excluded. An affirmative response would obviously imply that the objective should be included. A negative response would be taken as sufficient reason for exclusion. It is important when using this method to avoid excluding a large set of attributes, each of which fails the test of importance but which collectively are important. As the decision-making process proceeds and further insight is gained, the test of importance can be repeated with the excluded objectives to assure that they remain non-determinative. Otherwise they must be added to the hierarchy and evaluated for further decomposition themselves until new stopping points are reached.

The decomposition of objectives stops when the set of performance objectives is operationally useful and quantifiable, and the decision-maker, in consultation with appropriate stakeholders, is satisfied that it captures the expectations contained in the top-level objective. It is desirable that the performance objectives have the following properties. They should be:

- Complete – The set of performance objectives is complete if it includes all areas of concern embedded in the top-level objective.

- Operational – The performance objectives must be meaningful to the decision-maker so that he or she can understand the implications of meeting or not meeting them to various degrees. The decision-maker must ultimately be able to articulate a rationale for preferring one decision alternative over all others, which requires that he or she be able to ascribe value, at least qualitatively, to the degree to which the various alternatives meet the performance objectives.

- Non-redundant – The set of performance objectives is non-redundant if no objective contains, or significantly overlaps with, another objective. This is not to say that the ability of a particular alternative to meet different performance objectives will not be correlated. For example, in application, *maximize reliability* is often negatively correlated with *minimize cost*. Rather, performance objectives should be conceptually distinct, regardless of any solution-specific performance dependencies.

- Solution independent – The set of performance objectives should be applicable to any reasonable decision alternative and should not presuppose any particular aspect of an alternative to the exclusion of other reasonable alternatives. For example, an objectives hierarchy for a payload launch capability that had *Minimize Slag Formation* as a performance objective would be presupposing a solid propellant design. Unless solid propellant was specifically required based on a prior higher-level decision, *Minimize Slag Formation* would not reflect an unbiased decomposition of the top-level objective.

Guidance on developing objectives hierarchies can be found in Clemen [14] and Keeney and Raiffa [15], as well as on websites such as Comparative Risk Assessment Framework and Tools (CRAFT) [18].

Fundamental vs. Means Objectives

When developing an objectives hierarchy it is important to use *fundamental objectives* as opposed to *means objectives*. Fundamental objectives represent *what* one wishes to accomplish, as opposed to means objectives, which represent *how* one might accomplish it. Objectives hierarchies decompose high-level fundamental objectives into their constituent parts (partitioning), such that the fundamental objectives at the lower level are those that are implied by the fundamental objective at the higher level. In contrast, means objectives indicate a particular way of accomplishing a higher-level objective. Assessment of decision alternatives in terms of fundamental objectives as opposed to means objectives represents a performance-based approach to decision making, as recommended by the Aerospace Safety Advisory Panel (ASAP) as emphasizing "early risk identification to guide design, thus enabling creative design approaches that might be more efficient, safer, or both." [19]

The difference between fundamental objectives and means objectives is illustrated in Figure 16, which shows an objectives hierarchy on the top and a means objectives network on the bottom. The first thing to notice is that the objectives hierarchy is just that, a hierarchy. Each level decomposes the previous level into a more detailed statement of what the objectives entail. The objective, *Maximize Safety*, is decomposed (by partitioning) into *Minimize Loss of Life*, *Minimize Serious Injuries*, and *Minimize Minor Injuries*. The three performance objectives explain what is meant by *Maximize Safety*, without presupposing a particular way of doing so.[8]

In contrast, the means objectives network is not a decomposition of objectives, which is why it is structured as a network instead of a hierarchy. The objective, *Educate Public about Safety*, does not explain what is meant by any one the higher-level objectives; instead, it is a way of accomplishing them. Other ways may be equally effective or even more so. Deterministic standards in general are means objectives, as they typically prescribe techniques and practices by which fundamental objectives, such as safety, will be achieved. Means objectives networks arise in another context in the RIDM process and are discussed further in Section 3.2.1.

Performance Measures

Once an objectives hierarchy is completed that decomposes the top-level objective into a complete set of quantifiable performance objectives, a performance measure is assigned to each as the metric by which its degree of fulfillment is quantified. In many, if not most cases the appropriate performance measure to use is self-evident from the objective. In other cases the choice may not be as clear, and work must be done in order to assure that the objective is not only quantifiable, but that the performance measure used to quantify it is adequately representative of the objective.

---

[8] NASA is currently developing quantitative *safety goals* and associated *thresholds* (akin to imposed constraints) that will be used to guide risk acceptance decisions. [20] An example of a quantitative safety goal would be: the probability of loss of crew (P(LOC)) during the ascent phase of a crewed launch to LEO should be less than <a specified value>.

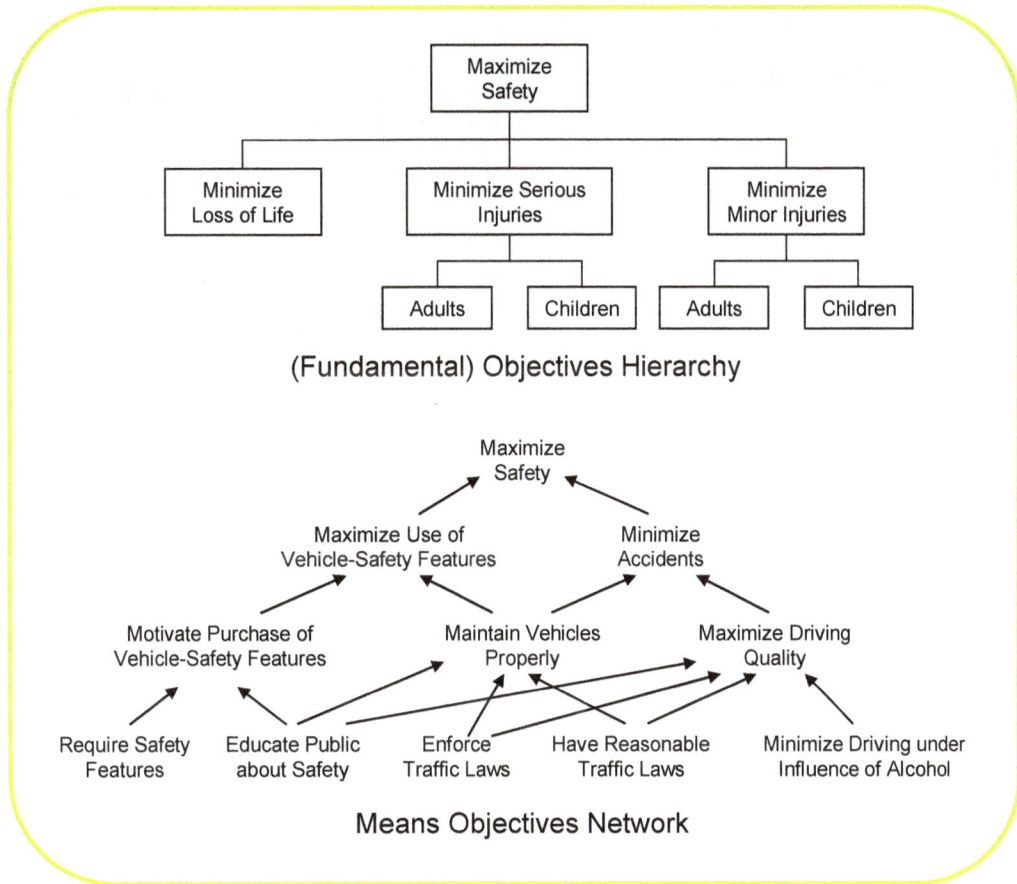

Figure 16. Fundamental vs. Means Objectives [21]

Objectives that have natural unit scales (e.g., *Minimize Cost*, *Maximize Payload*) are generally easy to associate with an appropriate performance measures (e.g., *Total Cost* or *Cost Overrun* [$], *Payload Mass* [kg]). Other objectives might not have an obvious or practical natural unit scale, thereby requiring the development of either a *constructed scale* or a *proxy performance measure*.

A constructed scale is typically appropriate for measuring objectives that are essentially subjective in character, or for which subjective or linguistic assessment is most appropriate. An example of such an objective might be *Maximize Stakeholder Support*. Here, stakeholder support is the attribute being measured, but there is no natural measurement scale by which an objective assessment of stakeholder support can be made. Instead, it might be reasonable to construct a scale that supports subjective/linguistic assessment of stakeholder support (see Table 1). Constructed scales are also useful as a means of quantifying what is essentially qualitative information, thereby allowing it to be integrated into a quantitative risk analysis framework.

Table 1. A Constructed Scale for Stakeholder Support (Adapted from [14])

Scale	Value	Description
5	Action-oriented Support	Two or more stakeholders are actively advocating and no stakeholders are opposed.
4	Support	No stakeholders are opposed and at least one stakeholder has expressed support.
3	Neutrality	All stakeholders are indifferent or uninterested.
2	Opposition	One or more stakeholders have expressed opposition, although no stakeholder is actively opposing.
1	Action-oriented Opposition	One or more stakeholders are actively opposing.

Alternatively, it may be possible to identify an objective performance measure that *indirectly* measures the degree of fulfillment of an objective. In the previous paragraph the objective, *Maximize Stakeholder Support*, was assessed subjectively using a *Stakeholder Support* performance measure with a constructed scale. Another strategy for assessing the objective might be to define a proxy for stakeholder support, such as the average number of stakeholders attending the bi-weekly status meetings. In this case, the proxy performance measure gives an indication of stakeholder support that might be operationally adequate for the decision at hand, although it does not necessarily correlate exactly to actual stakeholder support.

The relationships among natural, constructed and proxy scales are illustrated in Figure 17 in terms of whether or not the performance measure directly or indirectly represents the corresponding objective, and whether the assessment is empirically quantifiable or must be subjectively assessed. Additionally, Figure 17 highlights the following two characteristics of performance measures:

- The choice of performance measure type (natural, constructed, proxy) is not a function of the performance measure alone. It is also a function of the performance objective that the performance measure is intended to quantify. For example P(LOC) can be considered a natural performance measure as applied to astronaut life safety, since it directly addresses astronaut casualty expectation. However, in some situations P(LOC) might be a good *proxy* performance measure for overall astronaut health, particularly in situations where astronaut injury and/or illness are not directly assessable.

- There is seldom, if ever, a need for an indirect, subjective performance measure. This is because performance objectives tend to be intrinsically amenable to direct, subjective assessment. Thus, for objectives that do not have natural measurement scales, it is generally productive to ask whether the objective is better assessed directly but subjectively, or whether it is better to forego direct measurement in exchange for an empirically-quantifiable proxy performance measure. The first case leads to a constructed performance measure that is direct but perhaps not reproducible; the second to a performance measure that is reproducible but may not fully address the corresponding performance objective.

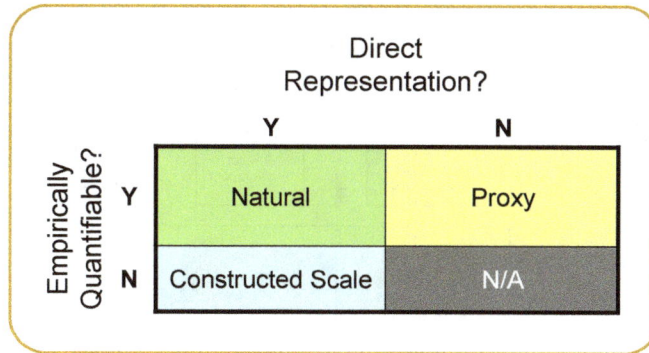

Figure 17. Types of Performance Measures

A performance measure should be adequate in indicating the degree to which the associated performance objective is met. This is generally not a problem for performance measures that have natural or constructed scales, but can be a challenge for proxy performance measures. In the *Maximize Stakeholder Support* example above, it is possible that a stakeholder who perceives the activity to be an obstacle to his or her real objectives might attend the meetings in order to remain informed about potential threats. Thus the average number of stakeholders attending the status meetings might not be an accurate representation of stakeholder support, and in this case may have a contraindicative element to it.

Figure 18 illustrates the relationship between performance objectives and performance measures. A performance measure has been established on each of the performance objectives based on the objective's natural measurement scale, a constructed scale that has been developed for subjective quantification, or via a proxy performance measure.

Although it is preferable that a performance measure be directly measurable, this is not always possible, even for objectives with natural measurement scales. For example, safety-related risk metrics such as *Probability of Loss of Mission, P(LOM)*, and *Probability of Loss of Crew, P(LOC)*, are typically used to quantify the objectives *Avoid Loss of Mission* and *Maintain Astronaut Health and Safety*. These performance measures are the product of modeling activities as opposed to direct measurement, involving the integration of numerous parameters within an analytical model of the alternative under consideration. In cases such as these, where modeling methods are integral to the resultant performance measure values, the modeling protocols become part of the performance measure definition. This assures that performance measures are calculated consistently.

One proxy performance measure of particular importance to many NASA decisions is *Flexibility*. *Flexibility* refers to the ability to support more than one current application. A technology choice that imposes a hard limit on the mass that can be boosted into a particular orbit has less flexibility than a choice that is more easily adaptable to boost more. The objective, *Maximize Flexibility*, allows this type of issue to be addressed systematically in decision making.

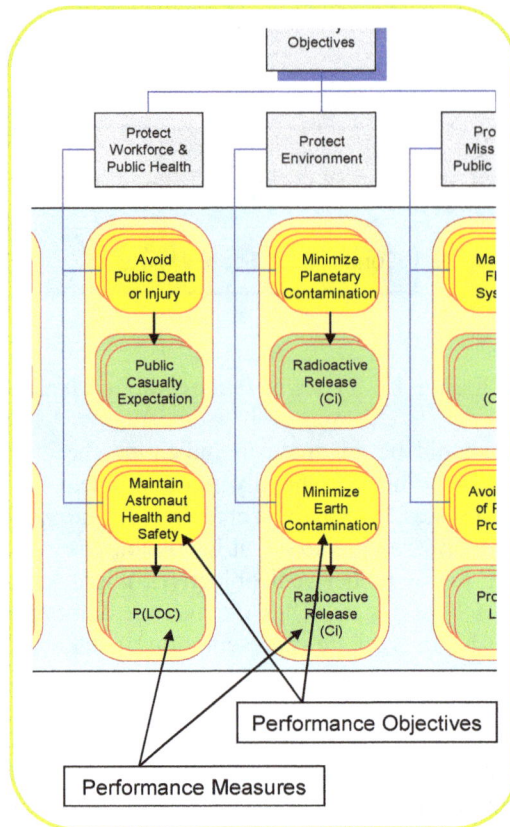

Figure 18. The Relationship between Performance Objectives and Performance Measures

However, since *Maximize Flexibility* refers to potential capabilities that are as yet undefined, there is no natural measurement scale that can be used for quantification.[9] A constructed scale is possible, although it requires subjective assessment. A proxy performance measure for flexibility can be constructed by, for example, assessing the capability of the alternative to support a selected set of alternative objectives, such as boosting a larger mass into orbit.

Risk Minimization Is Not a Performance Objective

It is sometimes the practice in decision analyses and trade studies to treat *Minimize Risk* as a distinct performance objective, which is then decomposed into domains such as technology, programmatic, cost, and schedule, resulting in performance measures such as *technology risk*, *programmatic risk, cost risk,* and *schedule risk*. However, in NPR 8000.4A, risk is the potential for shortfalls with respect to performance requirements (which in a RIDM context translates operationally into shortfalls with respect to performance commitments). Therefore, *Minimize Risk* is not a distinct objective in the objectives hierarchy. Rather, risk minimization is the task of

---

[9] In such applications, *Flexibility* is a surrogate for certain future performance attributes. This idea is discussed more extensively by Keeney [21] and Keeney and McDaniels [22].

risk management itself (including RIDM), for which risk is an attribute of every performance objective, as measured by the probability of falling short of its associated performance commitment.

For example, if a certain payload capability is contingent on the successful development of a particular propulsion technology, then the risk of not meeting the payload performance commitment is determined in part by the probability that the technology development program will be unsuccessful. In other words, the risk associated with technology development is accounted for in terms of its risk impact on the performance commitments (in this case, payload). There is no need to evaluate a separate Technology Risk metric.[10]

Example Performance Measures

Performance measures should fall within the mission execution domains of safety, technical, cost and schedule. Table 2 contains a list of typically important kinds of performance measures for planetary spacecraft and launch vehicles. Note that this is by no means a comprehensive and complete list. Although such lists can serve as checklists to assure comprehensiveness of the derived performance measure set, it must be stressed that performance measures are explicitly derived from top-level objectives in the context of stakeholder expectations, and cannot be established prescriptively from a predefined set.

Table 2. Performance Measures Examples for Planetary Spacecraft and Launch Vehicles

Performance Measures for Planetary Spacecraft	Performance Measures for Launch Vehicles
• End-of-mission (EOM) dry mass • Injected mass (includes EOM dry mass, baseline consumables and upper stage adaptor mass) • Consumables at EOM • Power demand (relative to supply) • Onboard data processing memory demand • Onboard data processing throughput time • Onboard data bus capacity • Total pointing error	• Total vehicle mass at launch • Payload mass (at nominal altitude or orbit) • Payload volume • Injection accuracy • Launch reliability • In-flight reliability • For reusable vehicles, percent of value recovered • For expendable vehicles, unit production cost at the nth unit

---

[10] Unless *Engage in Technology Development* is a performance objective in its own right.

Planetary Science Mission Example: Derive Performance Measures

From the generic top-level objective of "Project Success," the stakeholder expectations that have been captured are organized via an objectives hierarchy that decomposes the top-level objective through the mission execution domains of Safety, Technical, Cost, and Schedule, producing a set of performance objectives at the leaves. The terminology of "Minimize" and "Maximize" is used as appropriate to indicate the "direction of goodness" that corresponds to increasing performance for that objective.

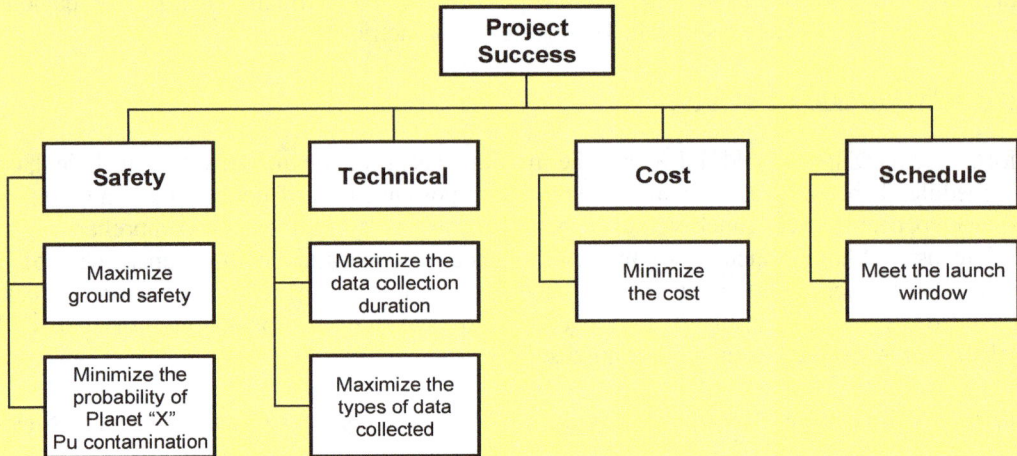

Objectives Hierarchy for the Planetary Science Mission Example

A quantitative performance measure is associated with each performance objective, along with any applicable imposed constraints. Below are the performance measures and applicable imposed constraints for four of the performance objectives. These are the four performance measures that will be quantified for the example in subsequent steps. In practice, all performance objectives are quantified.

**Selected Performance Measures and Imposed Constraints
for the Planetary Science Mission Example**

Performance Objective	Performance Measure	Imposed Constraint
Minimize Cost	Project cost ($M)	$500M
Minimize Development Time	Months to completion	55 months
Minimize the Probability of Planet "X" Pu Contamination	Probability of Planet "X" Pu Contamination	0.1%
Maximize Data Collection	Months of data collection	6 months

3.1.2 Step 2 - Compile Feasible Alternatives

The objective of Step 2 is to compile a comprehensive list of feasible decision alternatives through a discussion of a reasonable range of alternatives. The result is a set of alternatives that can potentially achieve objectives and warrant the investment of resources required to analyze them further.

3.1.2.1 Compiling an Initial Set of Alternatives

Decision alternatives developed under the design solution definition process [2] are the starting point. These may be revised, and unacceptable alternatives removed after deliberation by stakeholders based upon criteria such as violation of flight rules, violation of safety standards, etc. Any listing of alternatives will by its nature produce both practical and impractical alternatives. It would be of little use to seriously consider an alternative that cannot be adopted; nevertheless, the initial set of proposed alternatives should be conservatively broad in order to reduce the possibility of excluding potentially attractive alternatives from the outset. Keep in mind that novel solutions may provide a basis for the granting of exceptions and/or waivers from deterministic standards, if it can be shown that the intents of the standards are met, with confidence, by other means. In general, it is important to avoid limiting the range of proposed alternatives based on prejudgments or biases.

Defining feasible alternatives requires an understanding of the technologies available, or potentially available, at the time the system is needed. Each alternative should be documented qualitatively in a description sheet. The format of the description sheet should, at a minimum, clarify the allocation of required functions to that alternative's lower-level components. The discussion should also include alternatives which are capable of avoiding or substantially lessening any significant risks, even if these alternatives would be more costly. If an alternative would cause one or more significant risk(s) in addition to those already identified, the significant effects of the alternative should be discussed as part of the identification process.

Stakeholder involvement is necessary when compiling decision alternatives, to assure that legitimate ideas are considered and that no stakeholder feels unduly disenfranchised from the decision process. It is expected that interested parties will have their own ideas about what constitutes an optimal solution, so care should be taken to actively solicit input. However, the initial set of alternatives need not consider those that are purely speculative. The alternatives should be limited to those that are potentially fruitful.

3.1.2.2 Structuring Possible Alternatives (e.g., Trade Trees)

One way to represent decision alternatives under consideration is by a trade tree. Initially, the trade tree contains a number of high-level decision alternatives representing high-level differences in the strategies used to address objectives. The tree is then developed in greater detail by determining a general category of options that are applicable to each strategy. Trade tree development continues iteratively until the leaves of the tree contain alternatives that are well enough defined to allow quantitative evaluation via risk analysis (see Section 3.2).

Along the way, branches of the trade tree containing unattractive categories can be pruned, as it becomes evident that the alternatives contained therein are either *infeasible* (i.e., they are incapable of satisfying the imposed constraints) or categorically inferior to alternatives on other branches. An alternative that is inferior to some other alternative with respect to every performance measure is said to be *dominated* by the superior alternative. At this point in the RIDM process, assessment of performance is high-level, depending on simplified analysis and/or expert opinion, etc. When performance measure values are quantified, they are done so as point estimates, using a conservative approach to estimation in order to err on the side of inclusion rather than elimination.

Figure 19 presents an example launch vehicle trade tree from the Exploration Systems Architecture Study (ESAS) [23]. At each node of the tree the alternatives were evaluated for feasibility within the cost and schedule constraints of the study's ground rules and assumptions. Infeasible options were pruned (shown in red), focusing further analytical attention on the retained branches (shown in green). The key output of this step is a set of alternatives deemed to be worth the effort of analyzing with care. Alternatives in this set have two key properties:

- They do not violate imposed constraints

- They are not known to be dominated by other alternatives (i.e., there is no other alternative in the set that is superior in every way).

Alternatives found to violate either of these properties can be screened out.

Figure 19. Example Launch Vehicle Trade Tree from ESAS

Planetary Science Mission Example: Compile Feasible Alternatives

A trade tree approach was used to develop potential alternatives for the mission to Planet "X." As shown in the trade tree below, the three attributes that were traded were the orbital insertion method (propulsive deceleration vs. aerocapture), the science package (lighter, low-fidelity instrumentation vs. heavier, high-fidelity instrumentation), and the launch vehicle (small, medium, and large). However, initial estimates of payload mass indicated that there was only one appropriately matched launch vehicle option to each combination of insertion method and science package. Thus, eight of the twelve initial options were screened out as being "infeasible" (as indicated by the red X's), leaving four alternatives to be forwarded to risk analysis (alternatives 1 – 4).

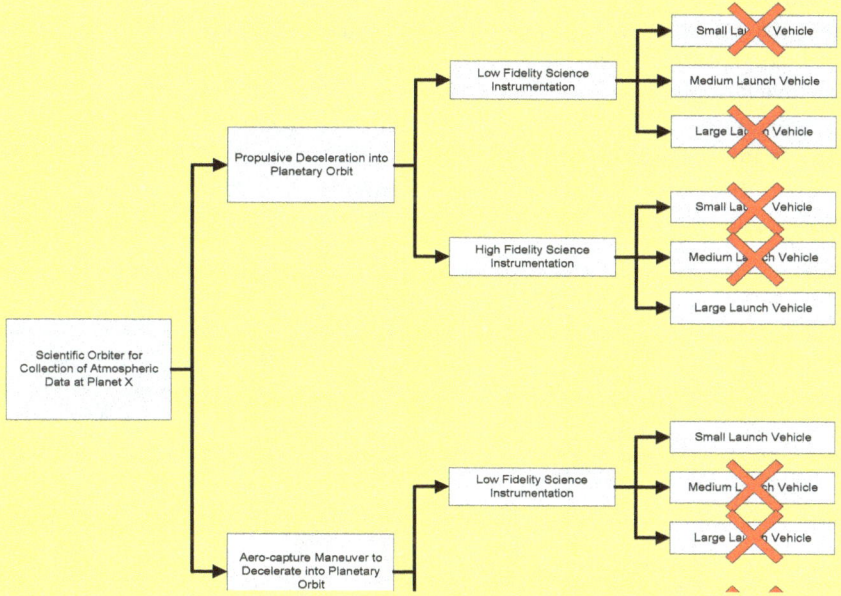

Trade Tree of Planetary Science Mission Alternatives

Feasible Alternatives forwarded to Risk Analysis

Alt #	Orbital Insertion Technology	Science Package	Launch Vehicle Size
1	Propulsive Insertion	Low Fidelity	Medium
2	Propulsive Insertion	High Fidelity	Large
3	Aerocapture	Low Fidelity	Small
4	Aerocapture	High Fidelity	Medium

3.2   Part 2 – Risk Analysis of Alternatives

Risk analysis consists of performance assessment supported by probabilistic modeling. It links the uncertainties inherent in a particular decision alternative to uncertainty in the achievement of objectives, were that decision alternative to be pursued. Performance is assessed in terms of the performance objectives developed in Step 1. The performance measures established for these objectives provide the means of quantifying performance so that alternatives can be effectively compared.

Figure 20 illustrates Part 2 of the RIDM process, Risk Analysis of Alternatives. In Step 3, risk analysis methodologies are selected for each analysis domain represented in the objectives, and coordination among the analysis activities is established to ensure a consistent, integrated evaluation of each alternative. In Step 4, the risk analysis is conducted, which entails probabilistic evaluation of each alternative's performance measure values, iterating the analysis at higher levels of resolution as needed to clearly distinguish performance among the alternatives. Then the TBfD is developed, which provides the primary means of risk-informing the subsequent selection process.

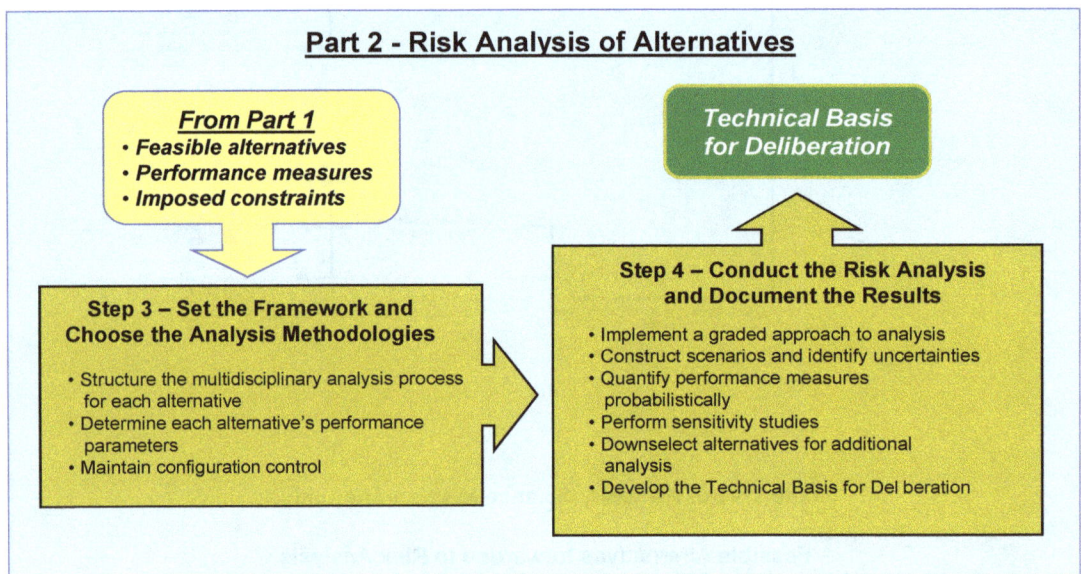

Figure 20. RIDM Process Part 2, Risk Analysis of Alternatives

3.2.1   Step 3 – Set the Framework and Choose the Analysis Methodologies

This step of the RIDM process is concerned with how domain-specific analyses, conducted in accordance with existing methodological practices, are integrated into a multidisciplinary framework to support decision making under uncertainty. In general, each mission execution domain has a suite of analysis methodologies available to it that range in cost, complexity, and time to execute, and which produce results that vary from highly uncertain rough order-of-magnitude (ROM) estimates to the detailed simulations. The challenge for the risk analysts is to establish a framework for analysis across mission execution domains that:

- Operates on a common set of (potentially uncertain) *performance parameters* for a given alternative (e.g., the cost model uses the same mass data as the lift capacity model)

- Consistently addresses uncertainties across mission execution domains and across alternatives (e.g., budget uncertainties, meteorological variability)

- Preserves correlations between performance measures (discussed further in Section 3.2.2)

- Is transparent and traceable.

The means by which a given level of performance will be achieved is alternative specific, and accordingly, the analyses that are required to support quantification are also alternative specific. For example, one alternative might meet the objective of *Minimize Crew Fatalities* by developing a high reliability system with high margins and liberal use of redundancy, eliminating the need for an abort capability. Since the high mass associated with the high margins of this approach impacts the objective, *Maximize Payload Capacity*, a different alternative might address the same crew safety objective by combining a lighter, less reliable system with an effective crew abort capability. For these two alternatives, significantly different analyses would need to be performed to quantify the probability *P(LOC)* of accomplishing the crew safety performance measure. In the first case, *P(LOC)* is directly related to system reliability. In the second case, reliability analysis plays a significant part, but additional analysis is needed to quantify abort effectiveness, which involves analysis of system responsiveness to the failure, and survivability given the failure environment.

Performance Parameters

A *performance parameter* is any value needed to execute the models that quantify the performance measures. Unlike performance measures, which are the same for all alternatives, performance parameters typically vary among alternatives, i.e., a performance parameter that is defined for one alternative might not apply to another alternative.

Example performance parameters related to the performance objective of lofting X lbs into low Earth orbit (LEO) might include propellant type, propellant mass, engine type/specifications, throttle level, etc. Additionally, performance parameters also include relevant environmental characteristics such as meteorological conditions.

Performance parameters may be uncertain. Indeed, risk has its origins in performance parameter uncertainty, which propagates through the risk model, resulting in performance measure uncertainty.

3.2.1.1 Structuring the Analysis Process

For a given alternative, the relationship between performance measures and the analyses needed to quantify them can be established and illustrated using a means objectives network (introduced in Section 3.1.1). Figure 21, adapted from [24], illustrates the idea. This figure traces Performance Parameter 1 through the risk analysis framework, showing how it is used by multiple risk analyses in multiple mission execution domains.

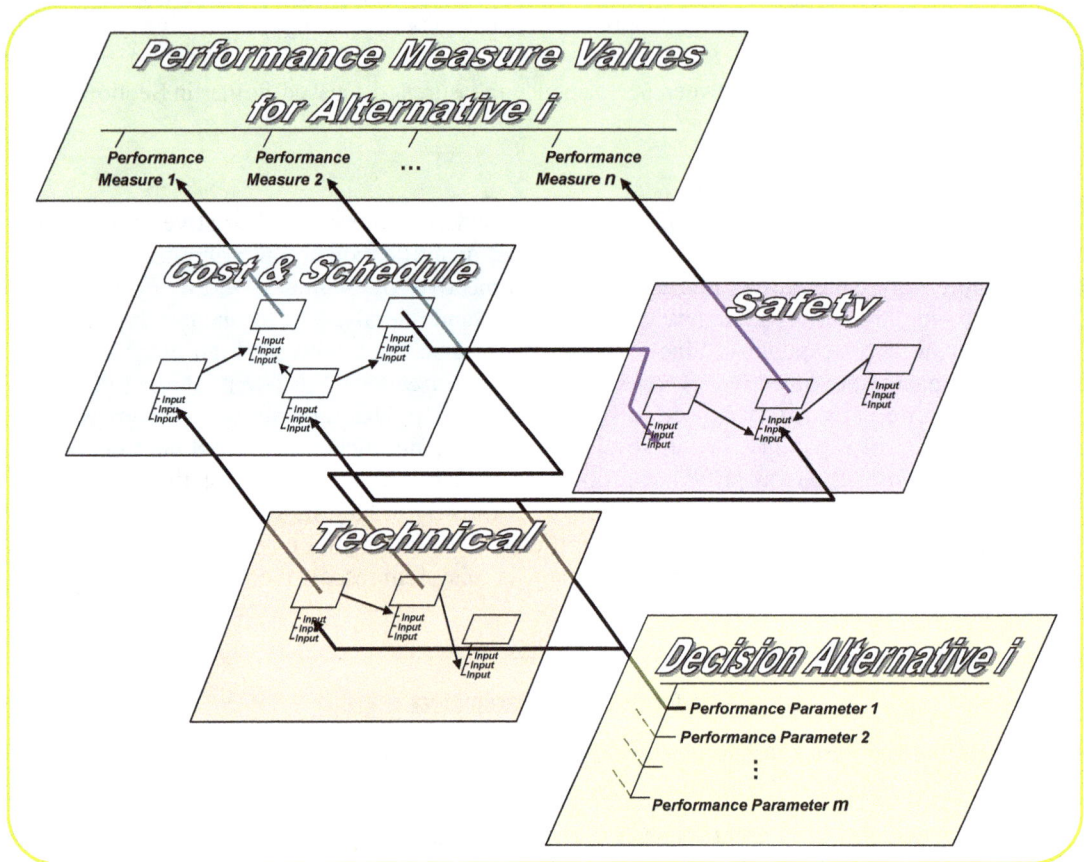

Figure 21. Risk Analysis Framework (Alternative Specific)

For example, Performance Parameter 1 is a direct input to a risk model in the Cost and Schedule mission execution domains (which have been combined in the figure for convenience). This analysis produces outputs that are used as inputs to two other Cost and Schedule risk models. One of these produces a value for Performance Measure 1, whereas the other produces an output that is needed by a risk model in the Safety mission execution domain. This Safety risk model ultimately supports quantification of Performance Measure **n**.

Each of the **m** performance parameters that defines Alternative *i* can be similarly traced through the risk analysis framework.

Figure 21 illustrates the need for coordination among the organizations conducting the analyses to assure that:

- There is an organization responsible for the quantification of each performance measure;

- The data requirements for every risk model are understood and the data sources and destinations have been identified;

- All data are traceable back through the risk analysis framework to the performance parameters of the analyzed alternative.

3.2.1.2 Configuration Control

It is important to maintain consistency over the definition of each analyzed alternative to ensure that all involved parties are working from a common data set. This is particularly true during the earlier phases of the program/project life cycle where designs may be evolving rapidly as decisions are made that narrow the trade space and extend it to higher levels of detail. It is also true when decisions are revisited, such as during requirements rebaselining (as discussed in Section 2.2.2), in which case the complete definition of the alternative may be distributed among various organizational units at different levels of the NASA hierarchy. In this case it is necessary for the organization at the level of the decision to consolidate all relevant alternative data at its own level, as well as levels below, into a configuration managed data set.

Additionally, the risk analysis framework itself must be configuration controlled, in terms of the analyses (e.g., version number) and data pathways.

3.2.1.3 Implementing Various Levels of Model Rigor in Selecting Risk Analysis Methods

The spectrum of analysis disciplines involved in the risk analysis of alternatives is as broad as the spectrum of performance measures, spanning the mission execution domains of Safety, Technical, Cost, and Schedule. It is not the intent of this handbook to provide detailed guidance on the conduct of domain-specific analyses. Such guidance is available in domain-specific documents like NPR 8715.3C, NASA General Safety Program Requirements [25] (Chapter 2, System Safety), the NASA Cost Estimating Handbook [26], the NASA Systems Engineering Handbook [2], and the NASA Probabilistic Risk Assessment Procedures Guide [27].

In this discussion, it is important to differentiate between model rigor and the concept of graded analysis. Model rigor (the subject of this section) pertains to the level of detail that is included in a model and is proportional to the maturity of the design. It would not make sense, for example, to include modeling of components in an integrated safety analysis if the components had not yet been specified in the design. On the other hand, graded analysis (the subject of the next section and later sections in Chapter 4) pertains more to the completeness of the modeling. In a graded analysis, the effort devoted to analyzing a particular risk issue or scenario is proportional to the importance of the issue or scenario being considered and whether it affects the ability to make an informed decision. It would not be necessary, for example, to analyze all the possible scenarios associated with an initiating event or all the uncertainties involved in the quantification of the resulting scenarios if the probability of the initiating event were vanishingly small.

Depending on project scale, life cycle phase, etc., different levels of model rigor are appropriate. As a general rule of thumb, the rigor of modeling and analysis should increase with successive program/project life cycle phases. In addition for a given phase, parametric, engineering, and logic modeling can commence at a low level of detail; the level of detail can be increased in an iterative fashion based on the requirement to reach a robust decision. Figure 22 indicates the types of analysis that are generally appropriate, as a function of life cycle phase, for cost, technical, and safety estimation. The figure is not meant to be prescriptive; there may be cases where detailed analysis of a phenomenon can provide guiding results early in the design process, before the specific details of a design are finalized.

Discussion of uncertainty can be found in Section 3.2.2. Detailed information on methods can be found in discipline-specific guidance, e.g., [2], [28], and [29].

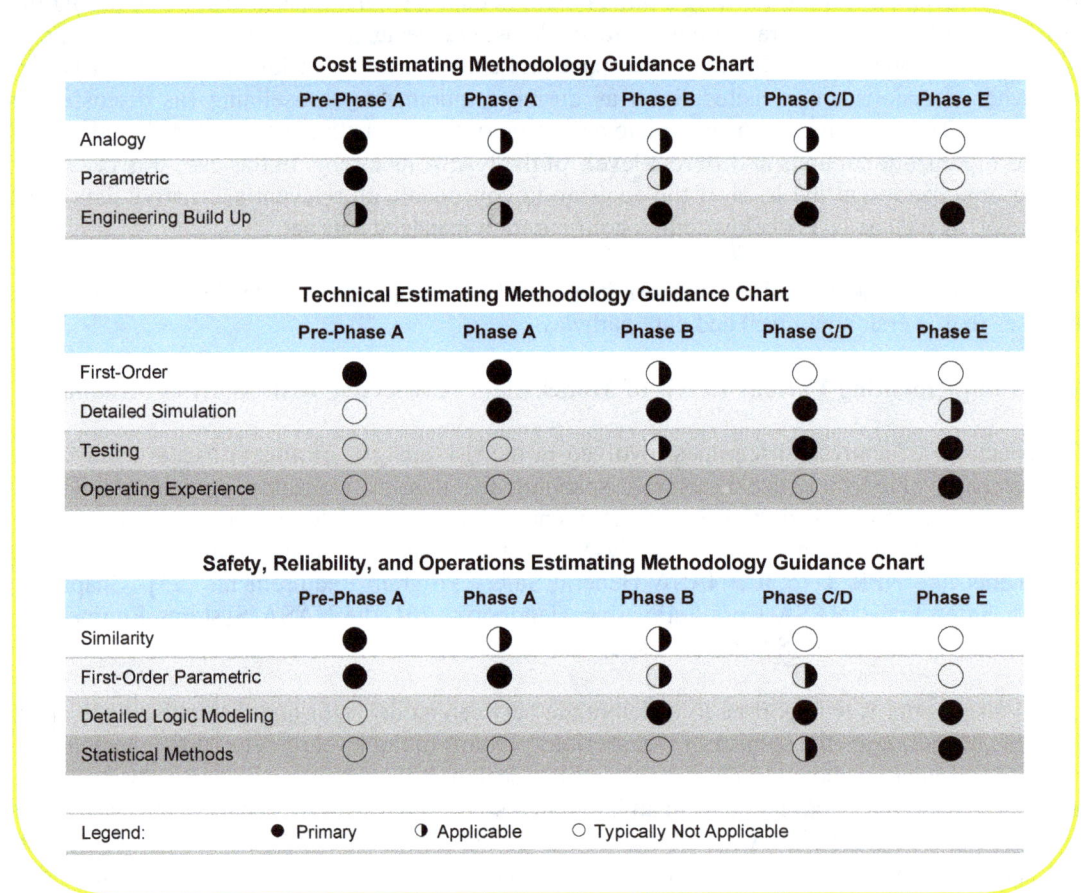

Cost Estimating Methodology Guidance Chart

	Pre-Phase A	Phase A	Phase B	Phase C/D	Phase E
Analogy	●	◐	◐	◐	○
Parametric	●	●	◐	◐	○
Engineering Build Up	◐	◐	●	●	●

Technical Estimating Methodology Guidance Chart

	Pre-Phase A	Phase A	Phase B	Phase C/D	Phase E
First-Order	●	●	◐	○	○
Detailed Simulation	○	●	●	●	◐
Testing	○	○	◐	●	●
Operating Experience	○	○	○	○	●

Safety, Reliability, and Operations Estimating Methodology Guidance Chart

	Pre-Phase A	Phase A	Phase B	Phase C/D	Phase E
Similarity	●	◐	◐	○	○
First-Order Parametric	●	●	◐	◐	○
Detailed Logic Modeling	○	○	●	●	●
Statistical Methods	○	○	○	◐	●

Legend: ● Primary   ◐ Applicable   ○ Typically Not Applicable

Figure 22. Analysis Methodology Guidance Chart

- **Cost and Schedule Estimating Methodologies:**

  - *Analogy Estimating Methodology* - Analogy estimates are performed on the basis of comparison and extrapolation to like items or efforts. Cost data from one past program that is technically representative of the program to be estimated serve as the basis of estimate. These data are then subjectively adjusted upward or downward, depending upon whether the subject system is believed to be more or less complex than the analogous program.

  - *Parametric Estimating* - Estimates created using a parametric approach are based on historical data and mathematical expressions relating cost as the dependent variable to selected, independent, cost-driving variables through regression analysis. Generally, an estimator selects parametric estimating when only a few key pieces of data are known, such as weight and volume. The implicit assumption of parametric estimating is that the same forces that affected cost in the past will affect cost in the future.

  - *Engineering Build Up Methodology* - Sometimes referred to as "grass roots" or "bottom-up" estimating, the engineering build up methodology rolls up individual estimates for each element into the overall estimate. This methodology involves the computation of the cost of a work breakdown structure (WBS) element by estimating at the lowest level of detail (often referred to as the "work package" level) wherein the resources to accomplish the work effort are readily distinguishable and discernable. Often the labor requirements are estimated separately from material requirements. Overhead factors for cost elements such as Other Direct Costs (ODCs), General and Administrative (G&A) expenses, materials burden, and fees are generally applied to the labor and materials costs to complete the estimate.

- **Estimating Methodologies for Technical Performance Measures:**

  - *First-Order Estimating Methodology* - First-order estimates involve the use of closed-form or simple differential equations which can be solved given appropriate bounding conditions and/or a desired outcome without the need for control-volume based computational methods. The equations may be standard physics equations of state or empirically-derived relationships from operation of similar systems or components.

  - *Detailed Simulation Estimating Methodology* - Estimates using a detailed simulation require the construction of a model that represents the physical states of interest in a virtual manner using control-volume based computational methods or methods of a similar nature. These simulations typically require systems and conditions to be modeled to a high-level of fidelity and the use of "meshes" or network diagrams to represent the system, its environment (either internal, external, or both), and/or processes acting on the system or environment. Examples are computational fluid dynamics (CFD) and finite-element modeling.

o *Testing Methodology* - Testing can encompass the use of table-top experiments all the way up to full-scale prototypes operated under real-world conditions. The objective of the test is to measure how the system or its constituent components may perform within actual mission conditions. Testing could be used for assessing the expected performance of competing concepts or for evaluating whether the system or components will meet flight specifications.

o *Operating Experience Methodology* - Once the system is deployed data gathered during operation can be analyzed to provide empirically accurate representations of how the system will respond to different conditions and how it will operate throughout its lifetime. This information can serve as the basis for applicable changes, such as software uploads or procedural changes, that may improve the overall performance of the system. Testing and detailed simulation may be combined with operating experience to extrapolate from known operating conditions.

- **Safety, Reliability, & Operations Estimating Methodologies:**

o *Similarity Estimating Methodology* - Similarity estimates are performed on the basis of comparison and extrapolation to like items or efforts. Reliability and operational data from a past program that is technically representative of the program to be estimated serves as the basis of estimate. Reliability and operational data are then subjectively adjusted upward or downward, depending upon whether the subject system is believed to be more or less complex than the analogous program.

o *First-Order Parametric Estimation* - Estimates created using a parametric approach are based on historical data and mathematical expressions relating safety, reliability, and/or operational estimates as the dependent variable to selected, independent, driving variables through either regression analysis or first-order technical equations (e.g., higher pressures increase the likelihood of tank rupture). Generally, an estimator selects parametric estimating when the system and its concept of operation are at the conceptual stage. The implicit assumption of parametric estimating is that the same factors that shaped the safety, reliability, and operability in the past will affect the system/components being assessed.

o *Detailed Logic Modeling Estimation* - Detailed logic modeling estimation involves "top-down" developed but "bottom-up" quantified scenario-based or discrete-event logic models that segregate the system or processes to be evaluated into discrete segments that are then quantified and mathematically integrated through Boolean logic to produce the top-level safety, reliability, or operational estimate. Detailed technical simulation and/or testing, as well as operational data, can be used to assist in developing pdfs for quantification of the model. Typical methods for developing such models may include the use of fault trees, influence diagrams, and/or event trees.

- o *Statistical Methods* - Statistical methods can applied to data collected during system/component testing or from system operation during an actual mission. This is useful for characterizing the demonstrated safety, reliability, or operability of the system. In addition, patterns in the data may be modeled in a way that accounts for randomness and uncertainty in the observations, and then serve as the basis for design or procedural changes that may improve the overall safety, reliability, or operability of the system. These methods are useful for answering yes/no questions about the data (hypothesis testing), describing associations within the data (correlation), modeling relationships within the data (regression), extrapolation, interpolation, or simply for data mining activities.

3.2.1.4 Implementing a Graded Approach in Quantifying Individual Scenarios

In addition to increasing model rigor with successive program/project life cycle phases and the level of design detail available, the level of completeness exercised in the analysis should increase with the importance of the scenario being evaluated. Regardless of the time during the life cycle, certain scenarios will not be as important as others in affecting the performance measures that can be achieved for a given alternative. Scenarios that can be shown to have very low likelihood of occurrence and/or very low impacts on all the mission execution domains do not have to be evaluated using as rigorous a simulation methodology as scenarios that are deemed to be more important, and they do not have to be subjected to a full-blown accounting of the uncertainties. A point-estimate analysis using reasonably conservative simulation models and input parameter values should be sufficient for the evaluation of such scenarios.

The subject of graded analysis is particularly relevant to the assessment of how individual risks accumulate to form aggregate risks. This subject will be taken up in more detail in Section 4.3.1 within the context of CRM.

3.2.1.5 Use of Existing Analyses

The RIDM process does not imply a need for a whole new set of analyses. In general, some of the necessary analyses will already be planned or implemented as part of the systems engineering, cost estimating, and safety and mission assurance (S&MA) activities. Risk analysis for RIDM should take maximum advantage of existing activities, while also influencing them as needed in order to produce results that address objectives, at an appropriate level of rigor to support robust decision making.

3.2.2   Step 4 – Conduct the Risk Analysis and Document the Results

Once the risk analysis framework is established and risk analysis methods determined, performance measures can be quantified. As discussed previously, however, this is just the start of an iterative process of successive analysis refinement driven by stakeholder and decision-maker needs (see Part 3 of the RIDM process).

3.2.2.1 Probabilistic Modeling of Performance

If there were no uncertainty, the question of performance assessment would be one of quantifying point value performance measures for each decision alternative. In the real world, however, uncertainty is unavoidable, and the consequences of selecting a particular decision alternative cannot be known with absolute precision. When the decision involves a course of action, there is uncertainty in the unfolding of events, however well planned, that can affect the achievement of objectives. Budgets can shift, overruns can occur, technology development activities can encounter unforeseen phenomena (and often do). Even when the outcome is realized, uncertainty will still remain. Reliability and safety cannot be known absolutely, given finite testing and operational experience. The limits of phenomenological variability in system performance can likewise not be known absolutely, nor can the range of conditions under which a system will have to operate. All this is especially true at NASA, which operates on the cutting edge of scientific understanding and technological capability.

For decision making under uncertainty, risk analysis is necessary, in which uncertainties in the values of each alternative's performance parameters are identified and propagated through the analysis to produce uncertain performance measures (see Figure 6 in Section 1.5.2). Moreover, since performance measures might not be independent, correlation must be considered. For example, given that labor tends to constitute a high fraction of the overall cost of many NASA activities, cost and schedule tend to be highly correlated. High costs tend to be associated with slipped schedules, whereas lower costs tend to be associated with on-time execution of the program/project plan.

One way to preserve correlations is to conduct all analysis within a common Monte Carlo "shell" that samples from the common set of uncertain performance parameters, propagates them through the suite of analyses, and collects the resulting performance measures as a vector of performance measure values [28]. As the Monte Carlo shell iterates, these performance measure vectors accumulate in accordance with the parent joint pdf that is defined over the entire set of performance measures. Figure 23 notionally illustrates the Monte Carlo sampling procedure as it would be applied to a single decision alternative (Decision Alternative *i*).

Uncertainties are distinguished by two categorical groups: aleatory and epistemic [29], [30]. Aleatory uncertainties are random or stochastic in nature and cannot be reduced by obtaining more knowledge through testing or analysis. Examples include:

- The room-temperature properties of the materials used in a specific vehicle.

- The scenario(s) that will occur on a particular flight.

In the first case, there is random variability caused by the fact that two different material samples will not have the same exact properties even though they are fabricated in the same manner. In the second case, knowing the mean failure rates for all the components with a high degree of certainty will not tell us which random failures, if any, will actually occur during a particular flight. On the other hand, epistemic uncertainties are not random in nature and can be reduced by obtaining more knowledge through testing and analysis.

Planetary Science Mission Example: Set the Analysis Framework

The figure below shows the risk analysis framework used to integrate the domain-specific analyses. Each alternative is characterized by its performance parameters, some of which are uncertain (shown in red text) and others of which have definite, known deterministic values (shown in black text). In order to calculate the performance measures previously selected for illustrative purposes, four separate performance models have been developed for radiological contamination, data collection, schedule, and cost. Some performance parameters, such as *spacecraft structure mass*, *launch reliability*, and *science package TRL*, are used in multiple models. Some models (e.g., the data collection model) produce outputs (e.g., *science package mass*) that are inputs to other models (e.g., the schedule model).

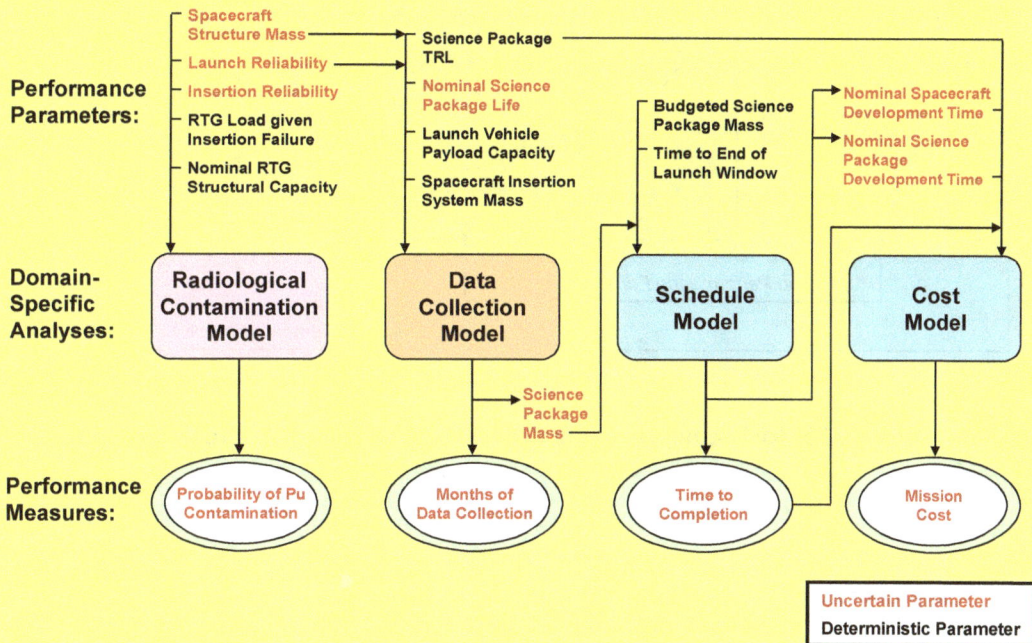

Risk Analysis Framework

The analysis framework shown above was used for all four alternatives selected for risk analysis. However, in general, each alternative may require its own analysis framework, which may differ substantially from other alternatives' frameworks based on physical or operational differences in the alternatives themselves. When this is the case, care should be taken to assure analytical consistency among alternatives in order to support valid comparisons.

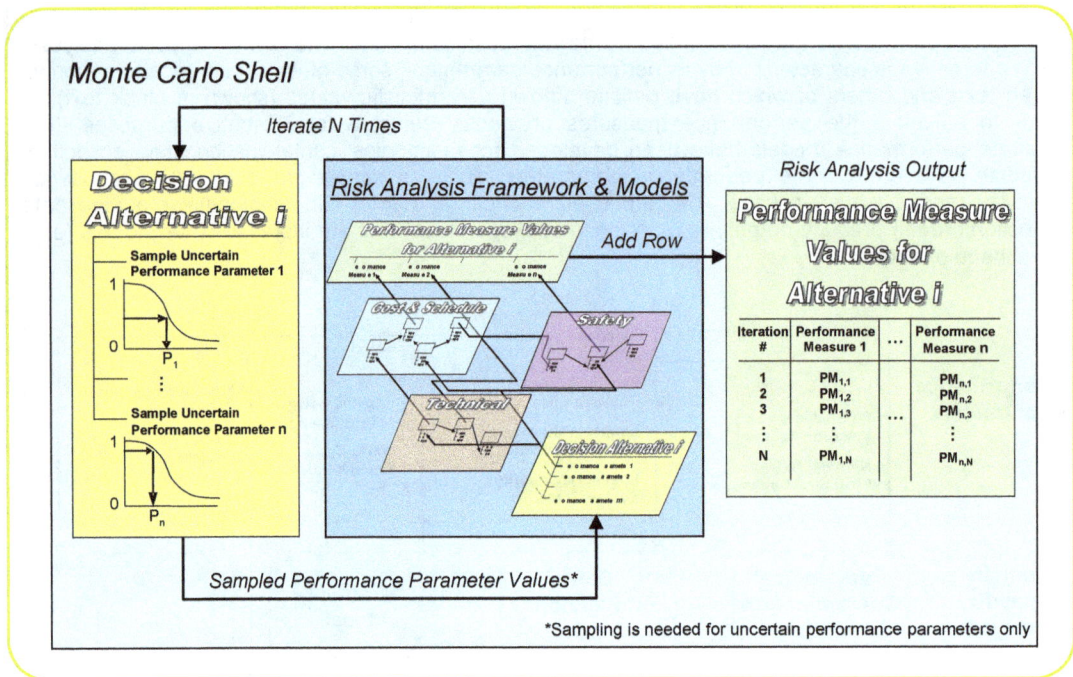

Figure 23. Risk Analysis Using a Monte Carlo Sampling Procedure

Examples include:

- The properties of a material at very high temperatures and pressures that are beyond the capability of an experimental apparatus to simulate

- The mean failure rates of new-technology components that have not been exhaustively tested to the point of failure in flight environments.

In both cases, the uncertainty is caused by missing or incomplete knowledge or by limitations in the models used to make predictions.

The assessed risk of not meeting performance requirements during a given mission are affected by both types of uncertainty. For example, if a component were required to operate for 17 years with 90% confidence during a flight to other planets in our solar system, and it had only been tested for 1 year, the evaluation of whether it meets the 90% confidence requirement would have to include both aleatory uncertainty (e.g., the possibility of a premature failure given a known mean failure rate) and epistemic uncertainty (e.g., uncertainty in the mean failure rate due to the limited test time). It is important to include both types of uncertainty in evaluating the performance risk. It is also important to know the relative contribution of each type of failure, since the former source of risk could not be reduced by more testing (without design modification) but the latter source could.

3.2.2.2 Use of Qualitative Information in RIDM

As discussed in the preceding section, uncertainties in the forecasted performance measures are caused by uncertainties in the input performance parameters and in the models that are used to calculate the outcomes. These parameter and modeling uncertainties may be expressed in either quantitative or qualitative terms. If a parameter is fundamentally quantitative in nature, it is represented as having an uncertainty distribution that is expressed in terms of numerical values. For example, the date that a part is delivered is a quantitative performance parameter because it is defined in terms of the number of days between a reference date (e.g., the project's initiation) and the delivery date. The date has a discrete numerical distribution because it changes in 24-hour increments. Most performance parameters, such as the cost of the part or its failure rate, have continuous numerical distributions.

A performance parameter can often also be expressed in terms of a constructed scale that is qualitative in nature. For example, the technology readiness level (TRL) at the time of project initiation is a qualitative parameter because it is defined in terms of ranks that are based on non-numerical information. A TRL of 1, for example, is defined by terms such as: "basic principles observed and reported," "transition from scientific research to applied research," "essential characteristics and behaviors of systems and architectures," "descriptive tools are mathematical formulations or algorithms." Such terms are not amenable to quantitative analysis without a significant amount of interpretation on the part of the analysts.

While the performance parameter may be either quantitative or qualitative, the probability scale for the uncertainty distribution of the performance parameter is generally defined in a quantitative manner. The probability scale may be either continuous or discrete (although in most cases it is continuous). For example, a five-tiered discretization of probabilities on a logarithmic scale might be based on binning the probabilities into the following ranges: 10^{-5} to 10^{-4} for level 1, 10^{-4} to 10^{-3} for level 2, 10^{-3} to 10^{-2} for level 3, 10^{-2} to 10^{-1} for level 4, and above 10^{-1} for level 5. It could be argued that the probability levels could also be defined in verbal terms such as "very unlikely to happen," "moderately likely to happen," and "very likely to happen." While these definitions are not numerical as stated, it is usually possible to ascertain the numerical ranges that the analyst has in mind when making these assignments. Thus, the probability should be relatable to a quantitative scale.

Various types of quantitative and qualitative uncertainty distributions for the input parameters and conditions are shown in Figure 24. Three of these (the top left and right charts and the lower right chart within the first bracket) are types of probability density functions, whereas the fourth chart (lower left) is a form of a complementary cumulative distribution function (CCDF). Either form of distribution (density form or cumulative form) may be used to express uncertainty. The choice is governed by whichever is the easier to construct, based on the content of the uncertainty information.

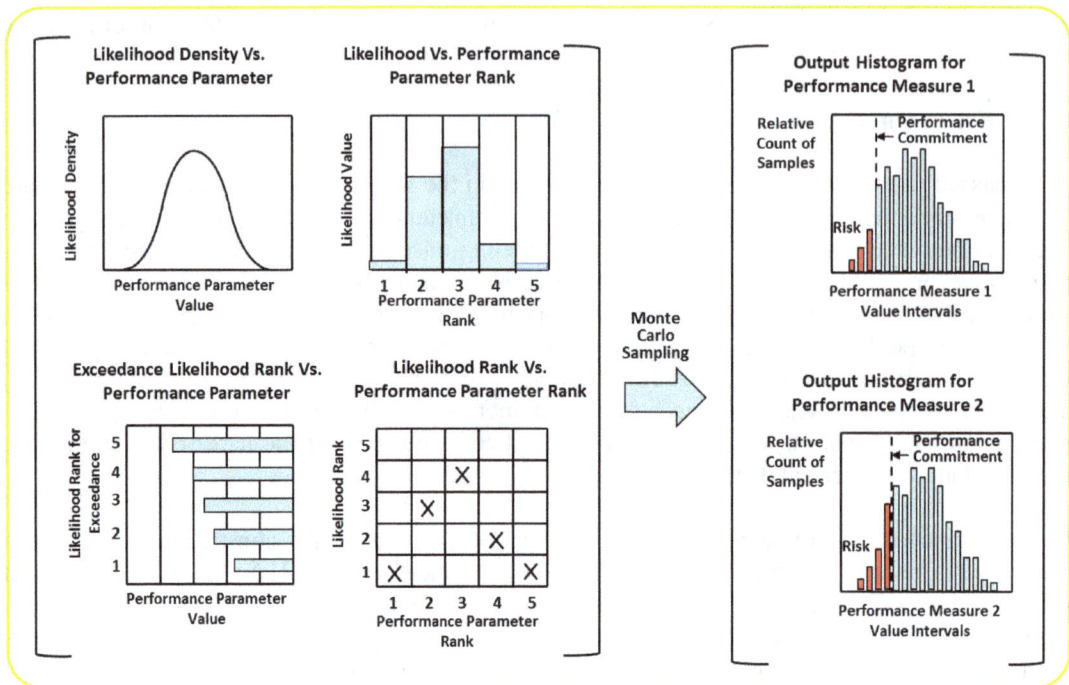

Figure 24. Uncertain Performance Parameters Leading to Performance Measure Histograms

It may be remarked that in cases where uncertainty distributions for performance parameters are derived from expert judgment elicitation, the corresponding pdf's may appear to violate intuition. For example, if two experts have fundamentally different viewpoints about the most likely value of an uncertain parameter, the best pdf accounting for that difference of opinion might be bimodal in shape: i.e., possessing two peaks. This might be the case, for example, if one expert believed that a new technology is likely to succeed and another believed the opposite. The pdf for likelihood of success might exhibit a peak at a high probability value, representing the former viewpoint, and another peak at a low probability, representing the latter viewpoint. The unusual shape of the pdf is not indicative of an error in analysis but rather a justifiable expression of differences between two viewpoints that are both reasonable given the evidence at hand. Differences of expert opinion in the formulation of uncertainty distributions will be taken up further in Section 3.2.2.5.

As depicted in Figure 24, the values of the output performance measures, as opposed to the values of the input performance parameters, are always quantitative in that they are defined in terms of numerical metrics. The output uncertainty distributions are expressed in the form of a histogram representation of output values obtained from Monte Carlo sampling of the input values and conditions.

Because the numerically based models are set up to accept numerical inputs, execution of the models for calculating the output performance measures is in general easier if all the performance parameters are defined in terms of quantitative scales, whether continuous or discrete. Caution should be used where one or more of the inputs are defined in terms of a qualitative, or constructed, scale. In these cases, the calculation of the performance measures may require that different models be used depending on the rank of the qualitative input. For example, the initial TRL for an engine might depend upon whether it can be made out of aluminum or has to be made out of beryllium. In this case, an aluminum engine has a higher TRL than a beryllium engine because the former is considered a heritage engine and the latter a developmental engine. On the other hand, a beryllium engine has the potential for higher thrust because it can run at higher temperatures. The model for calculating performance measures such as engine start-up reliability, peak thrust, launch date, and project cost would likely be different for an aluminum engine than for a beryllium engine.

The issue of how to use qualitative information within the RIDM process is especially important in dealing with institutional risks, where a greater percentage of performance requirements tend to be expressed in more qualitative terms than for program/project risks. Examples pertaining to the use of qualitative information in the evaluation of institutional risks will be taken up in an addendum to this handbook.

3.2.2.3 Risk Analysis Support of Robust Decision Making

Because the purpose of risk analysis in RIDM is to support decision making, the adequacy of the analysis methods must be determined in that context. The goal is a robust decision, where the decision-maker is confident that the selected decision alternative is actually the best one, given the state of knowledge at the time. This requires the risk analysis to be rigorous enough to discriminate between alternatives, especially for those performance measures that are determinative to the decision.

Figure 25 illustrates two hypothetical situations, both of which involve a decision situation having just one performance measure of significance. The graph on the left side of the figure shows a situation where Alternative 2 is clearly better than Alternative 1 (assuming that the pdfs are independent of each other) because the bulk of its pdf is to the left of Alternative 1's pdf. Thus the decision to select Alternative 2 is robust because there is high probability that a random sample from Alternative 1's pdf would perform better than a random sample from Alternative 2's pdf. In contrast, the graph on the right side of the figure shows a situation where the mean value of Alternative 1's performance measure is better than the mean value of Alternative 2's, but their pdfs overlap to a degree that prevents the decision to select Alternative 1 from being robust; that is, unless the pdfs for Alternatives 1 and 2 are highly correlated, there is a significant probability that Alternative 2 is actually better. The issue of correlated pdfs will be taken up later in this section.

For decisions involving multiple objectives and performance measures, it is not always possible to identify *a priori* which measures will be determinative to the decision and which will only be marginally influential. It is possible that some performance measures would require extensive analysis in order to distinguish between alternatives, even though the distinction would ultimately not be material to the decision. Consequently, the need for additional analysis for the purpose of making such distinctions comes from the deliberators and the decision-maker, as they deliberate the merits and drawbacks of the alternatives. The judgment of whether uncertainty reduction would clarify a distinction between contending decision alternatives is theirs to make; if it would be beneficial and if additional analysis is practical and effective towards that purpose, then the risk analysis is iterated and the results are updated accordingly.

Figure 25. Robustness and Uncertainty

3.2.2.4 Sequential Analysis and Downselection

While the ultimate selection of any given alternative rests squarely with the decision maker, he or she may delegate preliminary downselection authority to a local proxy decision-maker, in order to reduce the number of contending alternatives as early as practical in the decision-making process. There is no formula for downselection; it is an art whose practice benefits from experience. In general it is prudent to continuously screen the alternatives throughout the process. It is important to document the basis for eliminating such alternatives from further consideration at the time they are eliminated. Two such bases that were discussed in Section 3.1.2 are infeasibility and dominance. Additional discussion of downselection is presented in Section 3.3.2.

Downselection often involves the conduct of sequential analyses, each of which is followed by a pruning of alternatives. In this way, alternatives that are clearly unfavorable due to their performance on one (or few) performance measures can be eliminated from further analysis once those values are quantified.

For example, a lunar transportation architecture with Earth orbit rendezvous will require some level of loiter capability for the element(s) that are launched first (excluding simultaneous-launch options). A trade tree of architecture options might include a short loiter branch and a long loiter branch, corresponding to the times needed to span different numbers of trans-lunar injection (TLI) windows. It may not be known, prior to analysis, that the effects of propellant boil-off in terms of increased propellant needs, tankage, structure, and lift capacity, are prohibitive and render the long loiter option unfavorable. However, this circumstance can be determined based on an assessment of boil-off rate, and a sizing analysis for the architecture in the context of its DRMs. Once an analysis of sufficient rigor is performed, the entire long loiter branch of the trade tree can be pruned, without the need for additional, higher-level-of-rigor analyses on the potentially large number of long loiter alternatives compiled in Step 3 of the RIDM process.

Within the constraints of the analytical dependencies established by the risk analysis framework set in the previous step, it may be prudent to order the conduct of domain-specific analyses in a manner that exploits the potential for pruning alternatives prior to forwarding them for additional analysis. There is no hard rule for an optimal ordering; it depends on the specific decision being made, the alternatives compiled, and the analysis methods employed. It is recommended that opportunities for sequential analysis and downselection be looked for as alternatives are analyzed, and that the ordering of analyses be adjusted as appropriate to facilitate downselection, depending on which performance measures can be used as a basis for pruning (see Figure 26).

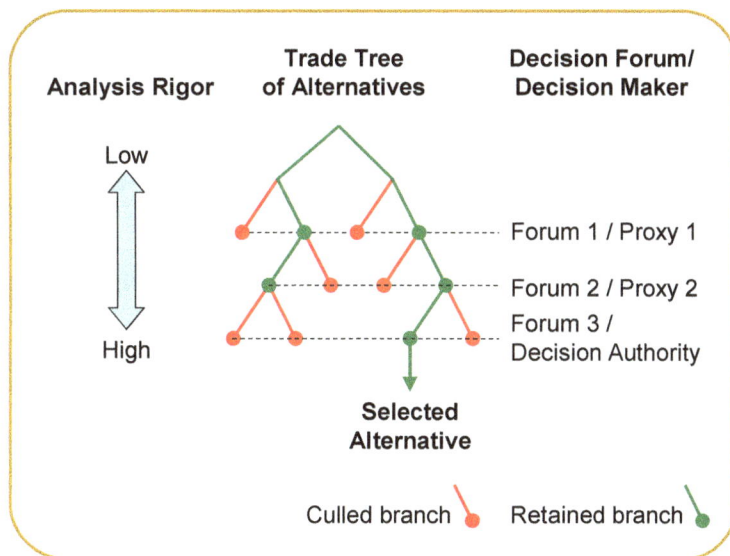

Figure 26. Downselection of Alternatives

Sequential analysis and downselection requires active collaboration among the risk analysts, the deliberators, and the decision maker. It is not the role of the risk analysts to eliminate alternatives except on the grounds of infeasibility. Sequential downselection, like all decision making, must be done in the context of stakeholder values and decision-maker responsibility/accountability.

Additionally, there is a potential vulnerability to sequential downselection, due to the incomplete quantification of performance measures it entails. It assumes that for the pruned alternatives, the level of performance in the analyzed domains is so poor that no level of performance in the unanalyzed domains could possibly make up for it. If this is not actually the case, an alternative that is attractive overall might be screened out due to inferior performance in just one particular area, despite superior performance in other domains and overall. Thus, it is good practice to review the validity of the downselects, in the context of the selected alternative, in order to assure that the selected alternative is indeed dominant.

3.2.2.5 Model Uncertainty and Sensitivity Studies

As is the case with all modeling activities, risk modeling typically entails a degree of model uncertainty to the extent that there is a lack of correspondence between the model and the alternative being modeled.

Many papers have been written on the subject of characterizing and quantifying model uncertainties; for example, surveys may be found in [31] and [32]. Often, the approaches advocate the use of expert judgment to formulate uncertainty distributions for the results from the models.

For example, suppose there was an existing model that produced a point value for thrust (a performance measure) based on a correlation of experimental data. In developing the correlation, the analysts emphasized the data points that produced lower thrust values over those that produced higher values in keeping with engineering practice to seek a realistically conservative result. In addition, the correlation was further biased to lower values to account for the fact that the experiments did not duplicate the high temperature, high pressure environment that is experienced during flight. A modified model was also derived that was similar to the original model but did not include any biasing of the data to produce a conservative result.

Based on this evidence, a set of subject matter experts made the following judgments:

- There is a 95% likelihood that the thrust during an actual flight will be higher than that predicted by the first model, which is known to be conservative.

- There is a 25% likelihood that the actual thrust will be higher than what is being predicted by the modified model, because the model does not introduce conservative assumptions in the data analysis, and in addition, the experimental simulation does not cover a range of environments where certain phenomena could decrease the thrust.

- There is only a 1% likelihood that the thrust will be lower than 0.8 times the values predicted by the original model, because there are no data to indicate that the thrust could be so low.

- There is a 1% likelihood that the thrust will be higher than 1.4 times the values predicted by the modified model because neither the original nor the modified model accounts for catalysis effects which could increase the thrust by up to 40%.

The analysts take this information to create a continuous distribution for the ratio of the actual thrust to that predicted by the original model (Figure 27). Thus, the modeling uncertainty for thrust is characterized by an adjustment factor that has a defined uncertainty distribution and is applied directly to the model output.

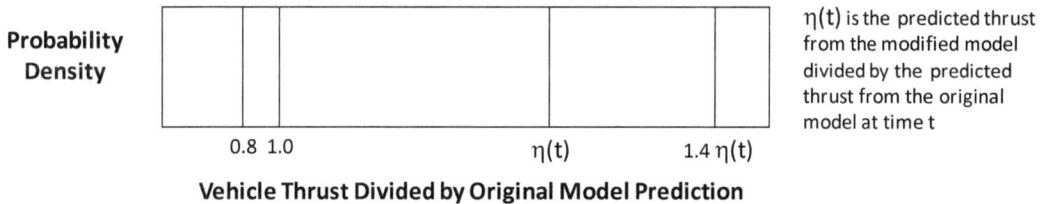

$\eta(t)$ is the predicted thrust from the modified model divided by the predicted thrust from the original model at time t

Probability Density

0.8 1.0 $\eta(t)$ 1.4 $\eta(t)$

Vehicle Thrust Divided by Original Model Prediction

Figure 27. Conceptualization of the Formulation of Modeling Uncertainty

The technique described above, where one or more models are subjectively assessed for bias as a means of quantifying model uncertainty, can in some sense be considered part of the overall modeling effort, as opposed to being a follow-on process that is applied to the risk analysis results. It is generically applicable in situations where subjective expertise, above and beyond that which is already captured in the available models, can be brought to bear to construct a "meta" model that integrates the information that is available from potentially diverse sources.

Another means of assuring that decisions are robust with respect to model uncertainty is to conduct sensitivity studies over ranges of credible model forms and/or parameter values. Sensitivity studies are particularly pertinent for models that produce point value performance measure results, even when the performance measure is known to be uncertain. In these cases, it is valuable to determine the sensitivity of the decision to bounding variations in the risk analysis assumptions. Figure 28 notionally presents the results of such a study. It shows how the preferred alternative varies as a function of assumptions about contractor support cost rate and payload mass. For example, if the contractor support cost rate is 120 and the payload mass is 18, then Alternative A is the preferred alternative. If, however, the assumed payload mass is 4, then Alternative B is preferable. More generally, if "Alternative B" is preferred for all reasonable values of contractor support cost rate and payload mass, then the decision is robust in favor of Alternative B (with respect to these parameters), without the need for additional rigor in determining the actual contractor support cost rate or payload mass. Likewise, if the reasonable range of these parameters falls entirely within the region "Alternative A," then the decision is robust for Alternative A. Only when the reasonable range of values straddles more than one region is more rigorous characterization of contractor support cost and payload mass needed for robust decision making.

3.2.2.6 Analysis Outputs

Like the variation in risk analysis methods, the analysis results presentation for RIDM may vary, depending on the nature of the problem being evaluated. Consequently, there can be no one standard analysis output. Instead, the results are tailored to the problem and the needs of the deliberation process. Consideration should be given for providing a variety of results, including:

- Scenario descriptions

- Performance measure pdfs and statistics

- Risk results (e.g., risk of not meeting imposed constraints)

- Uncertainty analyses and sensitivity studies

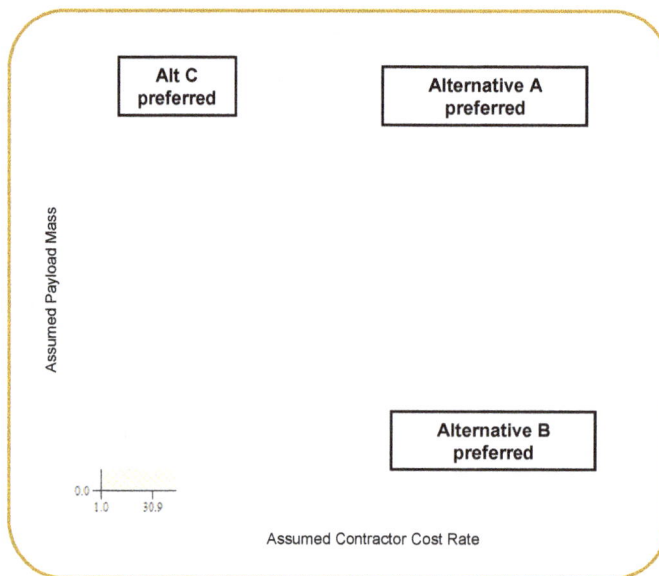

Figure 28. Notional Depiction of Decision Sensitivity to Input Parameters

It is important to note that the risk analysis results are expected to mature and evolve as the analysis iterates with the participation of the stakeholders and the decision-maker. This is not only due to increasing rigor of analysis as the stakeholders and the decision-maker strive for decision robustness. Additionally, as they establish firm performance commitments, it becomes possible to evaluate the analysis results in the context of those commitments. For example, prior to the development of performance commitments, it is not possible to construct a risk list that is keyed to the performance measures (except with respect to imposed constraints, which are firmly established prior to analysis).

3.2.2.7 Assessing the Credibility of the Risk Analysis Results

In a risk-informed decision environment, risk analysis is just one element of the decision-making process, and its influence on the decision is directly proportional to the regard in which it is held by the deliberators. A well-done risk analysis whose merits are underappreciated might not influence a decision significantly, resulting in a lost opportunity to use the available information to best advantage. Conversely, an inferior risk analysis held in overly high regard has the ability to produce poor decisions by distorting the perceived capabilities of the analyzed alternatives. In order to address this potential, an evaluation of the credibility of the risk analysis is warranted prior to deliberating the actual results.

NASA-STD-7009, Standard for Models and Simulations [33], provides the decision maker with an assessment of the modeling and simulation (M&S) results against key factors that:

- Contribute to a decision-maker's assessment of credibility

- Are sensibly assessed on a graduated credibility assessment scale (CAS).

Table 3 (which reproduces NASA-STD-7009 Table 1) presents a high-level summary of the evaluation criteria. These are explained in greater detail in the standard.
Table 3 by itself is not to be used in performing credibility assessments. Rather, the detailed level definitions in the standard are to be used.

Table 3. Key Aspects of Credibility Assessment Levels
(Factors with a Technical Review subfactor are underlined)

To assist in the application of the evaluation criteria dictated in NASA-STD-7009, Figure 29 presents a matrix indicating the "level" of analysis of each of the estimation methods.

Cost Estimating Method

Level	Analogy	Parametric	Engineering Build Up
0			
1	X		
2		X	
3			X
4			

Technical Estimating Method

Level	First-Order	Detailed Simulation	Testing	Operating Experience
0				
1	X			
2		X		
3			X	
4				X

Safety, Reliability, & Operations Estimating Method

Level	Similarity	First-Order Parametric	Detailed Logic Modeling	Statistical
0				
1	X			
2		X		
3			X	
4				X

Figure 29. Analysis Level Matrix

3.2.2.8 The Technical Basis for Deliberation

The TBfD (see Appendix C) specifies the minimum information needed to risk-inform the selection of a decision alternative. The content of the TBfD is driven by the question, "What information do the deliberators and decision-makers need in order for their decision process to be fully risk-informed?"

Graphical tools are recommended, in addition to tabular data, as a means of communicating risk results. At this point in the process, the imposed constraints are the only reference points with respect to which shortfalls can be determined, so they are the only things "at risk" so far. Figure 30 presents a notional color-coded chart of imposed constraint risk. In the figure, Alternative 7 is relatively low risk for every listed performance measure (i.e., those with imposed constraints on the allowable values), as well as for all constrained performance measures collectively (the "Total" column). Alternatives 12 and 3 have a mix of performance measure risks, some of which are high, resulting in a high risk of failing to meet one or more imposed constraints.

To assist the deliberators and decision-maker in focusing on the most promising alternatives, with an awareness of the relative risks to imposed constraints, the imposed constraints risk matrix has been:

- Sorted by total risk, with the least risky alternatives at the top

- Colored on a relative basis from low risk (the blue-violet end of the spectrum) to high risk (the orange-red end of the spectrum).

Alternative	Imposed Constraints					
	PM$_1$	PM$_2$	PM$_3$	...	PM$_n$	Total
	Constraint (> C$_1$)	Constraint (< C$_2$)	Constraint (> C$_3$)		Constraint (< C$_n$)	
7	0.7%*	0.09%	0.04%		0.1%	0.9%
1	2%	0.8%	0.2%		0.01%	4%
12	0.4%	5%	20%		0.1%	27%
...				...		
3	15%	0.9%	8%		0.01%	56%

*The probability of not meeting the imposed constraint

Figure 30. Notional Imposed Constraints Risk Matrix

When presenting the performance measure pdfs themselves, "band-aid" charts can be used, which show the mean, 5th percentile, and 95th percentile values (and often the median as well). Figure 31 shows a notional example of a band-aid chart. Unlike the imposed constraints matrix, which includes only those performance measures that have imposed constraints, band-aid charts can be made for every performance measure in the risk analysis, thereby giving a complete picture of the analyzed performance of each alternative.

When using charts such as the band-aid chart of Figure 31, it is important to know the degree of correlation among the different alternatives. For example, in the figure, the pdfs of Alternative 1 and Alternative 2 overlap to an extent that it may seem that the chances of either one having the higher performance are about the same. Indeed, this is true if the pdfs are independent. However, if they are correlated, then it might not be the case. For example, suppose the alternatives are identical except for some small design difference that slightly increases the value of Performance Measure X for Alternative 2. Then, although the performance of both alternatives is uncertain, the performance difference between them is known and constant.

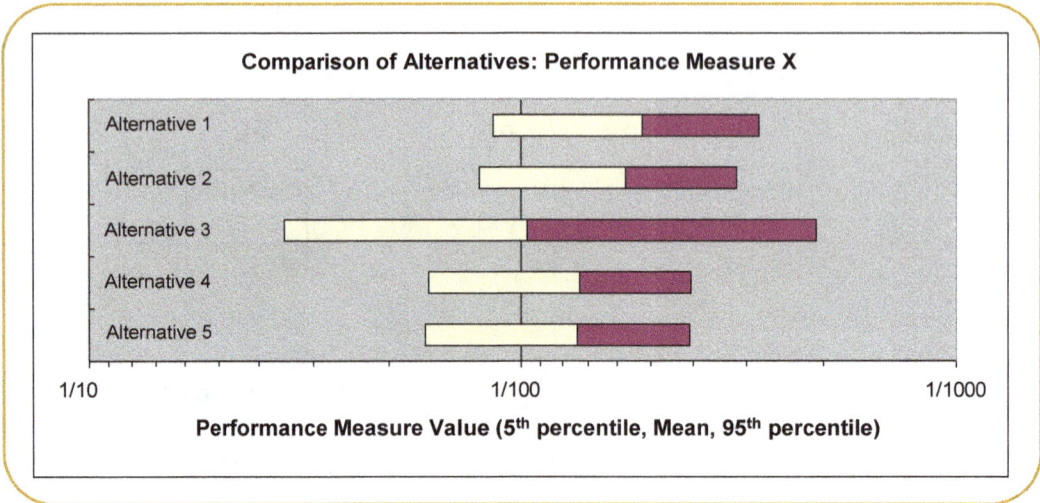

Figure 31. Notional Band Aid Chart for Performance Measure X

A direct representation of the difference between design alternatives, including the associated uncertainty, can supplement the information provided by band-aid charts, allowing for a better ability to make comparisons under uncertainty. A possible representation is shown in Figure 32 [34]. The figure shows performance measure pdfs for two alternatives whose performance measure values are correlated. A third, dotted, curve shows the pdf of the performance difference between the two alternatives. This curve indicates that despite the significant overlap between the two performance measure pdfs, Alternative 2 is unequivocally superior to Alternative 1, at least for the performance measure shown.

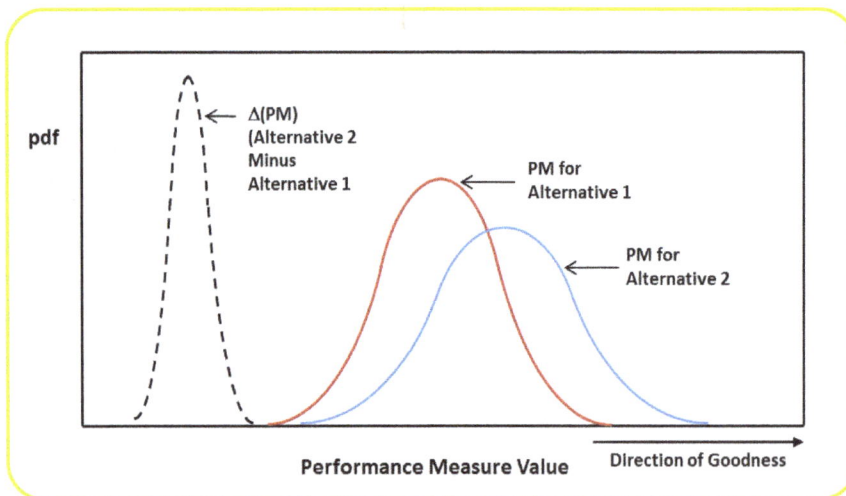

Figure 32. Comparison of Uncertainty Distributions

Planetary Science Mission Example: Conduct the Risk Analysis

Each of the modeled performance measures is quantified, using a Monte Carlo shell to sample the uncertain performance parameter pdfs and propagate them through the analysis framework, producing the performance measure results shown below. Each chart presents the assessed performance of the alternatives for a single performance measure, as well as the applicable imposed constraint, which defines the level of performance needed in order to fully meet the top-level objectives.

The risk analysis of alternatives produced the following pdfs:

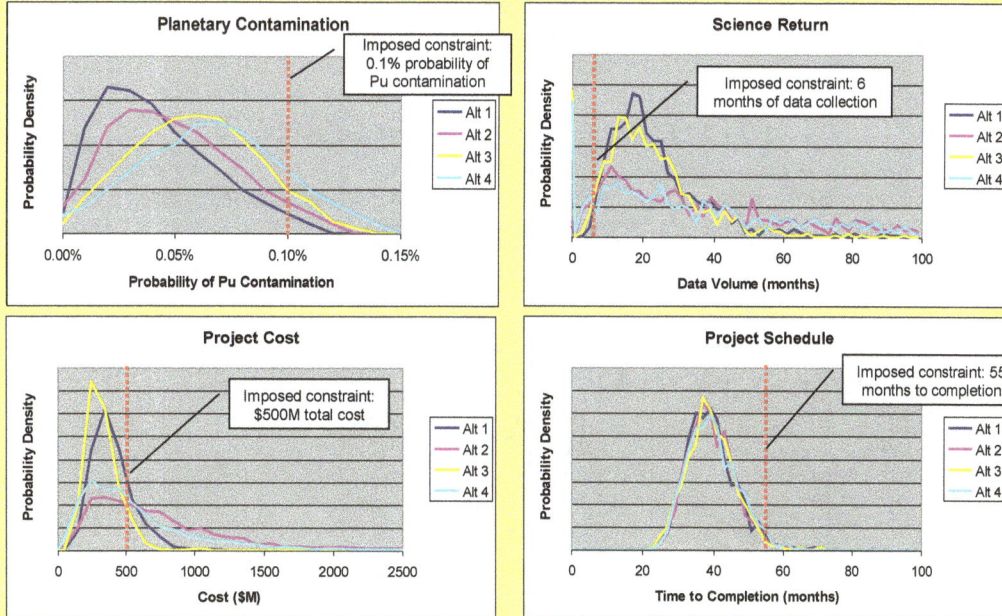

Performance Measure Results for the Planetary Science Mission Example

There is substantial overlap among the pdfs, particularly for *time to completion*. In practice, consideration is given to performing additional analysis to resolve such overlap in cases where doing so is expected to illuminate the decision. However, additional analysis might not help to distinguish among alternatives, especially when the underlying uncertainties are common to them. This is considered to be the case for the notional analysis of the Planetary Science Mission Example.

3.3   Part 3 – Risk-Informed Alternative Selection

The risk-informed alternative selection process within RIDM provides a method for integrating risk information into a deliberative process for decision making, relying on the judgment of the decision-makers to make a risk-informed decision. The decision-maker does not necessarily base his or her selection of a decision alternative solely on the results of the risk analysis. Rather, the risk analysis is just one input to the process, in recognition of the fact that it may not model everything of importance to the stakeholders. Deliberation employs critical thinking skills to the collective consideration of risk information, along with other issues of import to the stakeholders and the decision-maker, to support decision making.

Figure 33 illustrates Part 3 of the RIDM process, Risk-Informed Alternative Selection.

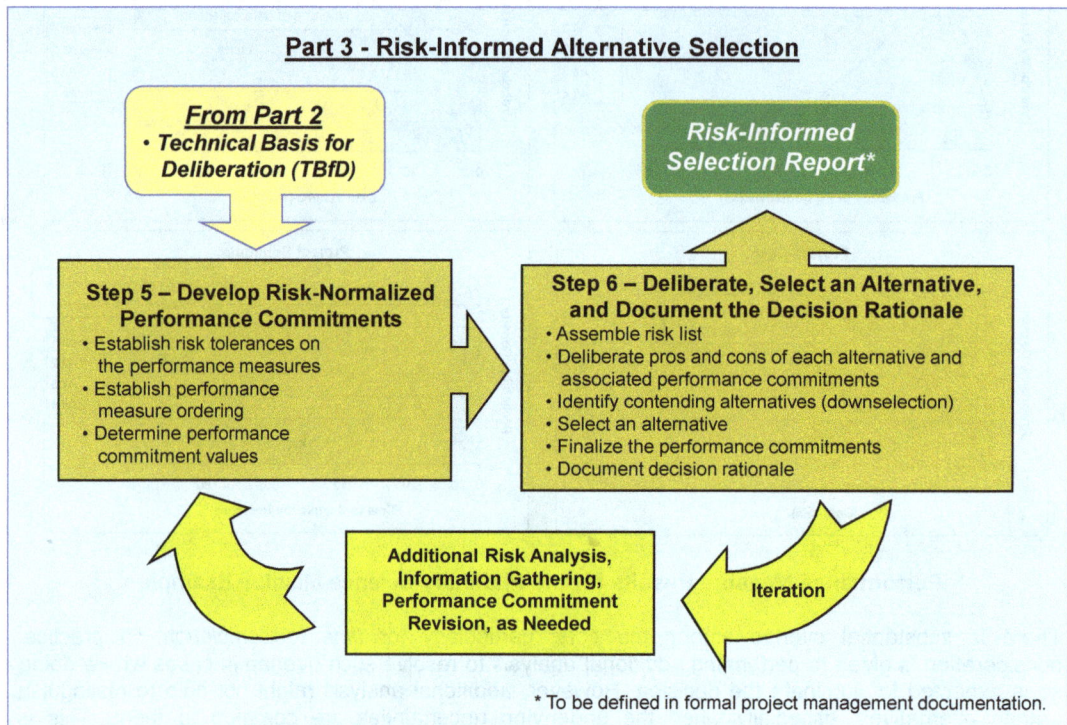

Figure 33. RIDM Process Part 3, Risk-Informed Alternative Selection

In Step 5, performance commitments are developed, representing consistent levels of risk tolerance across alternatives. In Step 6, relevant stakeholders, risk analysts, and decision-makers deliberate the relative merits and drawbacks of each alternative, given the information in the TBfD. This step is iterative, and may involve additional risk analysis or other information gathering as the participants strive to fully assess the alternatives and identify those that they consider to be reasonable contenders, worthy of serious consideration by the decision-maker. The decision-maker, or his/her proxy, may also be involved at this stage to help cull the number of alternatives (a.k.a. downselecting). Once a set of contending alternatives has been identified, the decision-maker integrates the issues raised during deliberation into a rationale for the selection of an alternative, and finalizes the performance commitments. The decision rationale is then documented in accordance with any existing project management directives, in a RISR. For pedagogical purposes, the process is laid out as if alternative selection involves a single decision that is made once deliberations are complete. However, as discussed in the Section 3.2.2.4 guidance on sequential analysis and downselection, decisions are often made in stages and in a number of forums that may involve a variety of proxy decision-makers.

Additionally, this handbook refers to the participants in deliberation as deliberators. This is also for pedagogical purposes. As illustrated in Figure 5, deliberators may be drawn from any of the sets of stakeholders, risk analysts, SMEs, and decision-makers.

3.3.1   Step 5 – Develop Risk-Normalized Performance Commitments

In order to generalize the consistent application of risk tolerance to the performance expectations of each decision alternative, this handbook introduces the concept of performance commitments. A performance commitment is a performance measure value set at a particular percentile of the performance measure's pdf, so as to anchor the decision-maker's perspective to that performance measure value as if it would be his/her commitment, were he/she to select that alternative. For a given performance measure, the performance commitment is set at the same percentile for all decision alternatives, so that the probability of failing to meet the different alternative commitment values is the same across alternatives.

Performance commitments support a *risk-normalized* comparison of decision alternatives, in that a uniform level of risk tolerance is established prior to deliberating the merits and drawbacks of the various alternatives. Put another way, risk-normalized performance commitments show what each alternative is capable of with an equal likelihood of achieving that capability, given the state of knowledge at the time.

The inputs to performance commitment development are:

- The performance measure pdfs for each decision alternative

- An ordering of the performance measures

- A risk tolerance for each performance measure, expressed as a percentile value.

For each alternative, performance commitments are established by sequentially determining, based on the performance measure ordering, the value that corresponds to the stated risk tolerance, conditional on meeting previously-defined performance commitments. This value becomes the performance commitment for the current performance measure, and the process is repeated until all performance commitments have been established for all performance measures. Figure 34 illustrates the process.

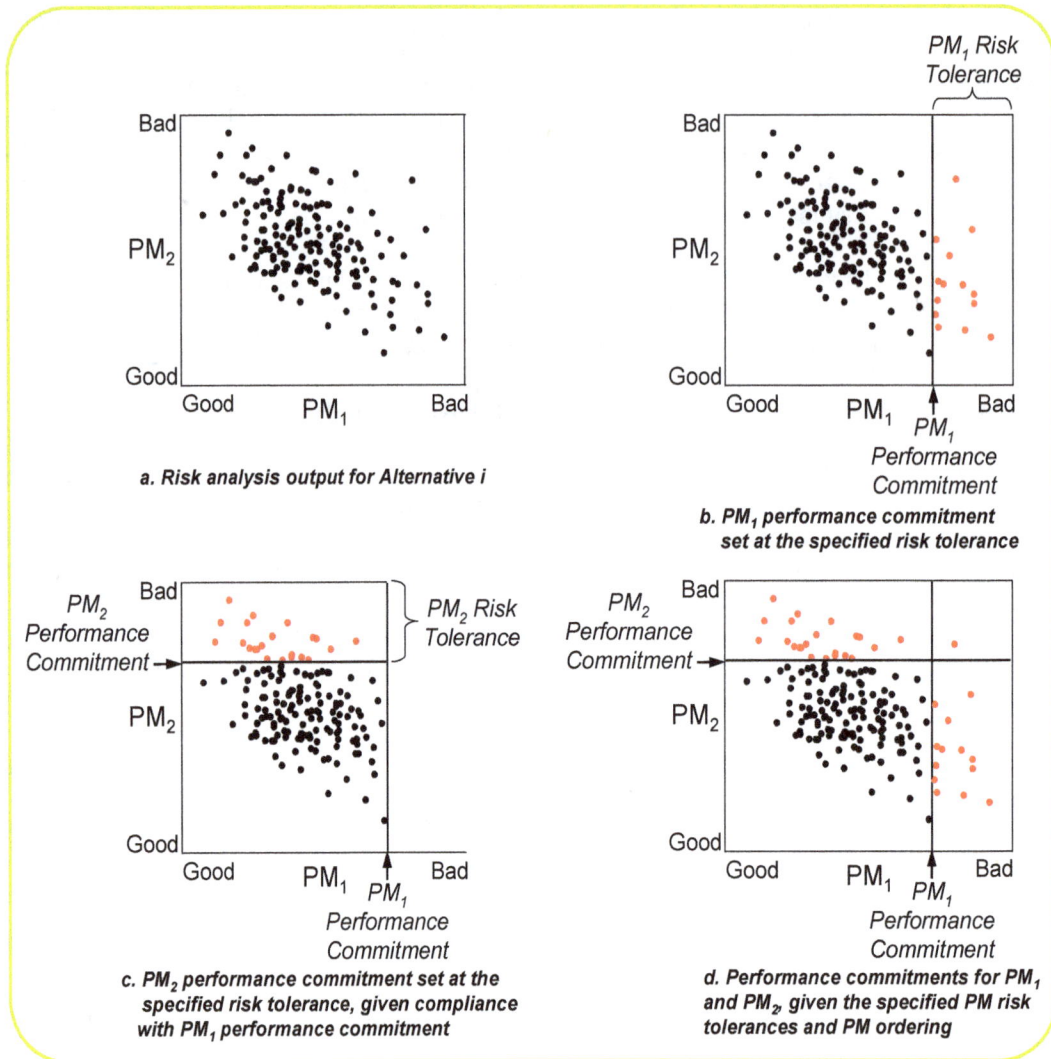

a. Risk analysis output for Alternative i

b. PM_1 performance commitment set at the specified risk tolerance

c. PM_2 performance commitment set at the specified risk tolerance, given compliance with PM_1 performance commitment

d. Performance commitments for PM_1 and PM_2, given the specified PM risk tolerances and PM ordering

Figure 34. Establishing Performance Commitments

In Figure 34 there are only two performance measures, PM_1 and PM_2. If, for example, PM_1 is P(LOC) and PM_2 is cost, then, the risk analysis results can be shown as a scatter plot on the P(LOC)-cost plane (see Figure 34a), where each point represents the output from a single iteration of the Monte Carlo shell. If the ordering of the performance measures is P(LOC) first and cost second, P(LOC) would be the first performance measure to have a performance commitment established for it (see Figure 34b). This is done by determining the value of P(LOC) whose probability of exceedance equals the defined risk tolerance. That value becomes the P(LOC) performance commitment.[11] The process is repeated for cost, *conditional on the P(LOC)*

---

[11] If the "direction of goodness" of the performance measure were reversed, the performance commitment would be at the value whose probability of exceedance equals one minus the risk tolerance.

performance commitment being met. Thus, the points on the scatter plot that exceed the P(LOC) performance commitment have been removed from consideration and the cost performance commitment is established solely on the basis of the remaining data (see Figure 34c). The result is a set of performance commitments for the P(LOC) and cost performance measures that reflects the risk tolerances of the deliberators and decision-maker (see Figure 34d). This procedure can be extended to any number of performance measures.

In general, different decision alternatives will have different performance commitments. But the probability of meeting each performance commitment will be the same (namely, one minus the risk tolerance of that performance measure), given that prior performance commitments in the performance measure ordering have been met:

$$\text{P(Performance Commitment i is met)} = 1 - \text{PM}_i \text{ Risk Tolerance}$$
$$= 1 - \text{P(Performance Commitment i is unmet | Performance Commitments j < i are met)}$$

Moreover, the probability of meeting all performance commitments is identical for all alternatives, and is calculated as:

P(All Pe

3.3.1.1 Establishing Risk Tolerances on the Performance Measures

The RIDM process calls for the specification of a risk tolerance for each performance measure, along with a performance measure ordering, as the basis for performance commitment development. These risk tolerance values have the following properties:

- The risk tolerance for a given performance measure is the same across all alternatives.

- Risk tolerance may vary across performance measures, in accordance with the stakeholders' and decision-maker's attitudes towards risk for each performance measure.

Risk tolerances, and their associated performance commitments, play multipurpose roles within the RIDM process:

- Uniform risk tolerance across alternatives normalizes project/program risk, enabling deliberations to take place that focus on performance capabilities on a risk-normalized basis.

- The risk tolerances that are established during the RIDM process indicate the levels of acceptable initial risk that the CRM process commits to managing during implementation. (Note: The *actual* initial risk is not established until performance requirements are agreed upon as part of the overall systems engineering process, and not explicitly addressed until the CRM process is initialized. More information on CRM initialization can be found in Section 4.)

- Performance commitments based on risk tolerance enable point value comparison of alternatives in a way that is appropriate to a situation that involves thresholds (e.g., imposed constraints). By comparing a performance commitment to a threshold, it is immediately clear whether or not the risk of crossing the threshold is within the established risk tolerance. In contrast, if a value such as the distribution mean were used to define performance commitments, the risk with respect to a given threshold would not be apparent.

Issues to consider when establishing risk tolerances include:

- *Relationship to imposed constraints* – In general, deliberators have a low tolerance for noncompliance with imposed constraints. Imposed constraints are akin to the success criteria for top-level objectives; if imposed constraints are not met, then objectives are not met and the endeavor fails. By establishing a correspondingly low risk tolerance on performance measures that have imposed constraints, stakeholders and decision-makers have assurance that if an alternative's performance commitments exceed the associated imposed constraints, there is a high likelihood of program/project success.

- *High-priority objectives* – It is expected that deliberators will also have a low risk tolerance for objectives that have high priority, but for which imposed constraints have not been set. The lack of an imposed constraint on a performance measure does not necessarily mean that the objective is of less importance; it may just mean that there is no well-defined threshold that defines success. This could be the case when dealing with quantities of data, sample return mass capabilities, or operational lifetimes. It is generally the case for life safety, for which it is difficult to establish a constraint *a priori*, but which is nevertheless always among NASA's top priorities.

- *Low-priority objectives and/or "stretch goals"* – Some decision situations might involve objectives that are not crucial to program/project success, but which provide an opportunity to take risks in an effort to achieve high performance. Technology development is often in this category, at least when removed from a project's critical path. In this case, a high risk tolerance could be appropriate, resulting in performance commitments that suggest the alternatives' performance potentials rather than their established capabilities.

- *Rebaselining issues* – Requirements on some performance measures might be seen as difficult to rebaseline. For these performance measures, deliberators might establish a low risk tolerance in order to reduce the possibility of having to rebaseline.

Risk tolerance values are up to the deliberators and decision maker, and are subject to adjustment as deliberation proceeds, opinions mature, and sensitivity excursions are explored. In particular, it is recommended that sensitivity excursions be explored over a reasonable range of risk tolerances, not only for the purpose of making a decision that is robust with respect to different risk tolerances, but also in order to find an appropriate balance between program/project risk and the performance that is specified by the performance commitments.

3.3.1.2 Ordering the Performance Measures

Because of possible correlations between performance measures, performance commitments are developed sequentially. As discussed in Section 3.3.1, performance commitments are defined at the value of a performance measure that corresponds to the defined risk tolerance, conditional on meeting previously defined performance commitments. In general, performance commitments depend on the order in which they are developed.

Qualitatively, the effect that performance measure order has on performance commitment values is a follows:

- If performance measures are independent, then the order is immaterial and the performance commitments will be set at the defined risk tolerances of the performance measures' marginal pdfs.

- If performance measures are positively correlated in terms of their directions of goodness, then the performance commitments that lag in the ordering will be set at higher levels of performance than would be suggested by their marginal pdfs alone. This is because lagging performance measures will have already been conditioned on good performance with respect to leading performance measures. This, in turn, will condition the lagging performance measures on good performance, too, due to the correlation.

- If performance measures are negatively correlated in terms of their directions of goodness, then the performance commitments that lag in the ordering will be set at lower levels of performance than would be suggested by their marginal pdfs alone. Figure 34 shows this phenomenon. In Figure 34c, the PM_2 performance commitment is set at a slightly lower performance than it would have been if the data points that exceed the PM_1 performance commitment were not "conditioned out."

- The lower the risk tolerance, the lower the effect of conditioning on subsequent performance commitments. This is simply because the quantity of data that is "conditioned out" is directly proportional to risk tolerance.

These general effects of performance measure ordering on performance commitments suggest the following ordering heuristics:

- Order performance measures from low risk tolerance to high risk tolerance. This assures a minimum of difference between the risk tolerances as defined on the conditioned pdfs versus the risk tolerances as applied to the marginal pdfs.

- Order performance measures in terms of the desire for specificity of the performance measure's risk tolerances. For example, the performance commitment for the first performance measure in the ordering is precisely at its marginal pdf. As subsequent performance commitments are set, dispersion can begin to accumulate as conditioning increases.

Once the performance commitments are developed, each alternative can be compared to every other alternative in terms of their performance commitments, with the deliberators understanding that the risk of not achieving the levels of performance given by the performance commitments is the same across alternatives. Additionally, the performance commitments can be compared to any imposed constraints to determine whether or not the possibility that they will not be satisfied is within the risk tolerance of the deliberators, and ultimately, the decision maker. Figure 35 notionally illustrates a set of performance commitments for each of three competing alternatives. Note that Alternative A does not satisfy the imposed constraint on payload capability within the risk tolerance that has been established for that performance measure.

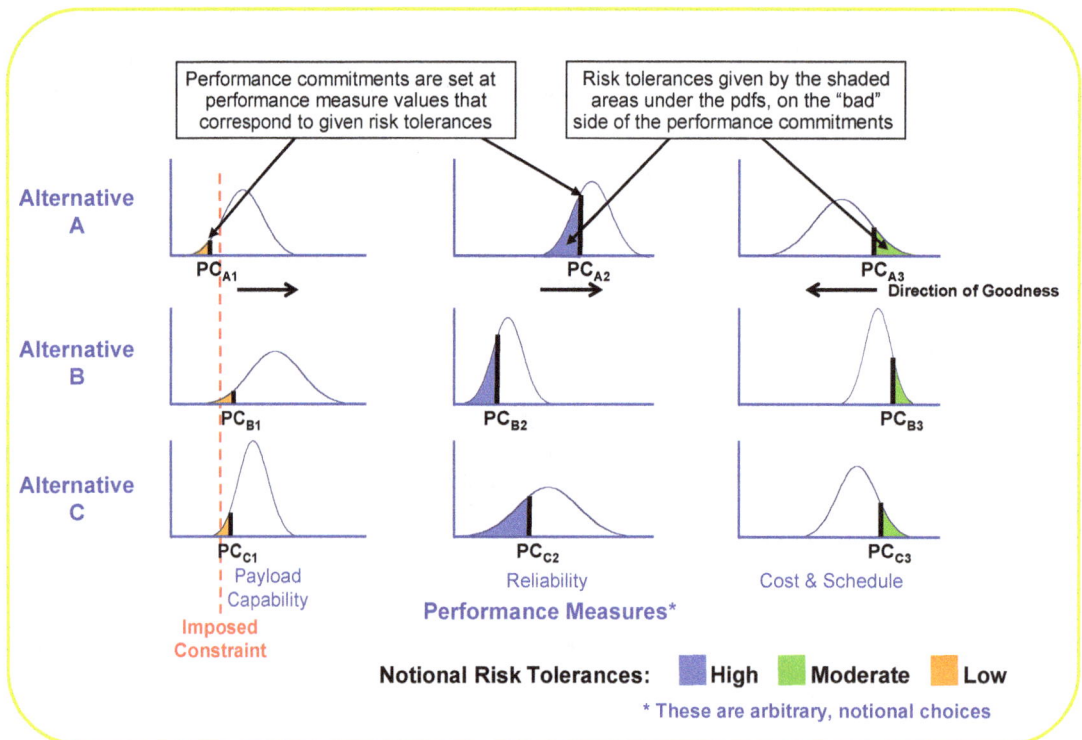

Figure 35. Performance Commitments and Risk Tolerances for Three Alternatives

Planetary Science Mission Example: Develop Risk-Normalized Performance Commitments

Risk-normalized performance commitments were developed for each of the four analyzed alternatives, for the performance measures of *time to completion*, *project cost*, *data volume*, and *planetary contamination*.

The table below shows the risk tolerance given to each. Because of the importance of meeting the 55 month launch window, a low risk tolerance of 3% is given to *time to completion*. Then, given the 3% *time to completion* risk tolerance, a 27% risk tolerance was given to *project cost* based on the NASA policy of budgeting cost and schedule at a joint confidence level (JCL) of 70% [7]. The 10% *data volume* risk tolerance is reasonably low, but reflects the belief that minor shortfalls will not significantly erode the success of the mission. The *planetary contamination* risk tolerance is moderately low, reflecting the importance of avoiding radiological releases into the planetary environment. These risk tolerance values are discussed with the decision maker to ensure that they are consistent with his/her views.

The table also shows the ordering of the performance measures that was used to develop performance commitments. *Time to completion* was chosen first due to its critical importance. *Project cost* was chosen next, due to its importance in an environment of scarce resources, and also because of its linkage to *time to completion* via the NASA JCL policy. Data volume was chosen third, due to its prominence among the technical objectives.

Performance Measure Risk Tolerance for
Planetary Science Mission Example

Performance Measure	Risk Tolerance	Performance Measure Ordering
Time to Completion	3%	1
Project Cost	27%	2
Data Volume	10%	3
Planetary Contamination	15%	4

The performance commitment chart (on the next page) shows the levels of performance that are achievable at the stated risk tolerances. One thing immediately evident is that Alternative 4 does not meet the 6 month *data volume* imposed constraint. However, the deliberation team recognizes that a different risk tolerance might produce a *data volume* performance commitment that is in line with the imposed constraint, so they are reluctant to simply discard Alternative 4 out of hand. Instead, they determine the risk that would have to be accepted in order to produce a *data volume* performance commitment of at least 6 months. This turns out to be 12%, which the team considers to be within the range of reasonable tolerances (indeed, it is not significantly different from 10%). In the interest of illuminating the situation to the decision maker, the team includes both sets of results in the chart.

Planetary Science Mission Example - Performance Commitment Chart

Risk Tolerance	Performance Commitments			
	Time to Completion (months)	Project Cost ($M)	Data Volume (months)	Planetary Contamination (probability)
Alternative	3%	27%	10% (12%)	15%
1. Propulsive Insertion, Low-Fidelity Science Package	54	476	11 (12)	0.07%
2. Propulsive Insertion, High-Fidelity Science Package	53	881	9.9 (11)	0.08%
3. Aerocapture, Low-Fidelity Science Package	55	413	6.8 (7.8)	0.10%
4. Aerocapture, High-Fidelity Science Package	54	688	4.9 (6.0)	0.11%

3.3.2   Step 6 – Deliberate, Select an Alternative, and Document the Decision Rationale

The RIDM process invests the decision-maker with the authority and responsibility for critical decisions. While ultimate responsibility for alternative selection rests with the decision-maker, alternative evaluation can be performed within a number of deliberation forums that may be held before the final selection is made. As partial decisions or "down-selects" may be made at any one of these deliberation forums, they are routinely structured around a team organizational structure identified by the decision-maker. It is important to have a team with broad based expertise to perform sufficient analysis to support a recommendation or decision. At the top of the structure may be the decision-maker or a deliberation lead appointed by the decision-maker. If a deliberation lead is appointed, this individual should be an experienced manager, preferably one with an analytical background.

3.3.2.1 Convening a Deliberation Forum

Deliberation forums address the major aspects of the decision. The use of these forums helps ensure that a responsible person leads each important area of analysis. The focus of these forums will vary with the type of study.

Depending on circumstances, forums can be split (e.g., into separate Safety and Technical), or functions can be combined (e.g., Cost and Schedule), or entirely new forums can be created (e.g., test, requirements, or stakeholder). The final choice of forum structure belongs to the decision-maker. At a minimum, the forums should mirror the major aspects of the study. Thus the creation of forums offers an important early opportunity to contemplate the effort's processes and goals. Every forum must have enough members to achieve a "critical mass" of knowledge, interest, and motivation. Typically, a small group with critical mass is more productive than a larger group with critical mass. This suggests starting with a small forum and adding members as necessary.

Members of a deliberation forum should ideally be selected based on their qualifications. Consideration should be given to those with relevant experience, knowledge, and interest in the subject matter. These individuals are frequently referred to as SMEs. In some cases they have an organizational charter to support the process and in other cases they participate because they are heavily invested in the outcome of the deliberation. When the most qualified are not available, the next most qualified should be sought.

People with diverse viewpoints on controversial issues should also be enlisted to participate in deliberations. They should represent the diversity of stakeholder interests. Partisans, by their nature, will defend their ideas and detect flaws in the ideas of their competition. This allows issues to be raised and resolved early that might otherwise lie in wait. A formal tracking system should be employed throughout the process to track items to closure.

Additional information on deliberative processes can be found in [17].

3.3.2.2 Identify Contending Alternatives

After the performance commitments have been generated, they are used to pare down the set of decision alternatives to those that are considered to be legitimate contenders for selection by the decision-maker. This part of the process is a continuation of the pruning activity begun in Section 3.1.2. At this point, however, the deliberators have the benefit of the TBfD and the performance commitments, as well as the subjective, values-based input of the deliberators themselves. Rationales for elimination of non-contending alternatives include:

- **Infeasibility** – Performance commitments are exceeded by the imposed constraints. In this case, imposed constraints cannot be met within the risk tolerance of the decision-maker.

- **Dominance** – Other alternatives exist that have superior performance commitments on every performance measure, and substantially superior performance on some.[12] In this case, an eliminated alternative may be feasible, but nonetheless is categorically inferior to one or more other alternatives.

- **Inferior Performance in Key Areas** – In general, in any decision involving multiple objectives, some objectives will be of greater importance to deliberators than others. Typically important objectives include crew safety, mission success, payload capability,

---

[12] When eliminating alternatives on the basis of dominance, it is prudent to allow some flexibility for uncertainty considerations beyond those captured by the performance commitments alone (discussed in the next subsection). Minor performance commitment shortfalls relative to other alternatives do not provide a strong rationale for elimination, absent a more detailed examination of performance uncertainty.

and data volume/quality. Alternatives that are markedly inferior in terms of their performance commitments in key areas can be eliminated on that basis, in recognition of stakeholder and decision-maker values.

Section 3.2.2.4 discusses sequential analysis and downselection, in which non-contending alternatives are identified and eliminated in parallel with risk analysis, thereby reducing the analysis burden imposed by the decision-making process. Sequential analysis and downselection represents a graded approach to the identification of contending alternatives, and is another example of the iterative and collaborative nature of the RIDM process.

3.3.2.3 Additional Uncertainty Considerations

The guidance above for identifying contending alternatives is primarily focused on comparisons of performance commitments. This facilitates comparisons between alternatives (and against imposed constraints), and the elimination of non-contenders from further consideration. However, performance commitments do not capture all potentially relevant aspects of performance, since they indicate the performance at only a single percentile of each performance measure pdf. Therefore, alternatives identified as contenders on the basis of their performance commitments are further evaluated on the basis of additional uncertainty considerations relating to their performance at other percentiles of their performance measure pdfs. In particular, performance uncertainty may give rise to alternatives with the following characteristics:

- **They offer superior expected performance** – In many decision contexts (specifically, those in which the decision-maker is *risk neutral*[13]), the decision-maker's preference for an alternative with uncertain performance is equivalent to his or her preference for an alternative that performs at the mean value of the performance measure pdf. When this is the case, expected performance is valuable input to decision making, as it reduces the comparison of performance among alternatives to a comparison of point values.

  However, in the presence of performance thresholds, over-reliance on expected performance in decision making has the potential to:

  o Introduce potentially significant probabilities of falling short of imposed constraints, thereby putting objectives at risk, even when the mean value meets the imposed constraints;

  o Contribute to the development of derived requirements that have a significant probability of not being achievable.

  Since direction-setting, requirements-producing decisions at NASA typically involve performance thresholds, expected performance should be considered in conjunction with performance commitments, to assure that the decision is properly risk informed.

---

[13] A risk-neutral decision maker is indifferent towards a decision between an alternative with a definite performance of X, versus an alternative having an uncertain performance whose mean value is X. In other words, a risk-neutral decision maker is neither disproportionately attracted to the possibility of exceptionally high performance (*risk seeking*) nor disproportionately averse to the possibility of exceptionally poor performance (*risk averse*).

- **They offer the potential for exceptionally high performance** – For a given performance measure pdf, the percentile value at the decision-maker's risk tolerance may be unexceptional relative to other contending alternatives. However, at higher risk tolerances, its performance may exceed that of other alternatives, to the extent that it becomes attractive relative to them. This may be the case even in the presence of inferior performance commitments on the same, or different, performance measures.

An example of this is shown notionally in Figure 36. In this figure, Alternative 2's performance commitment is at a worse level of performance than Alternative 1's; however, Alternative 2 offers a possibility of performance that is beyond the potential of Alternative 1. In this case, stakeholders and decision-makers have several choices. They can:

o Choose Alternative 1 on the basis of superior performance at their risk tolerance;

o Choose Alternative 2 on the basis that its performance at their risk tolerance, though not the best, is acceptable, and that it also has the potential for far superior performance; or

o Set their risk tolerance such that the performance commitment for both alternatives is the same thus making this performance measure a non-discriminator between the two options.

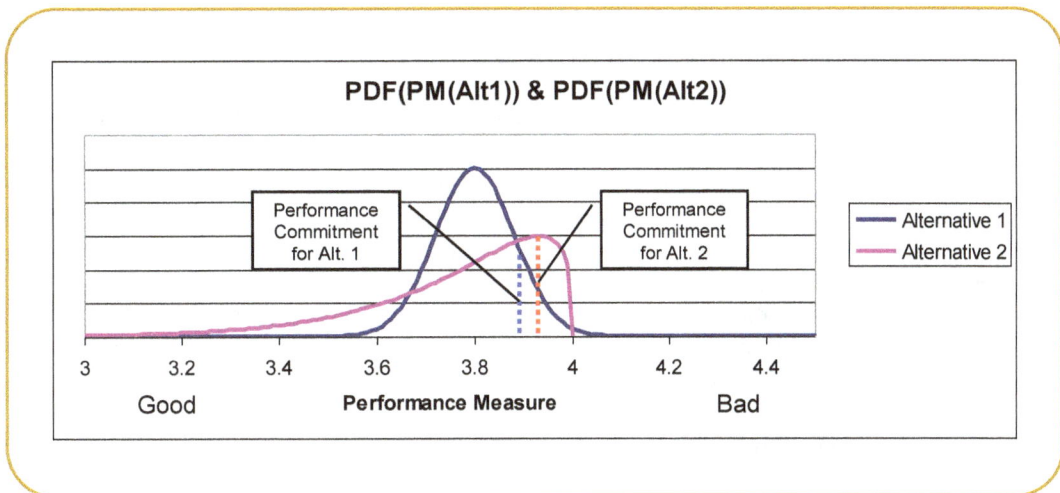

Figure 36. An Example Uncertainty Consideration: The Potential for High Performance

In the second case, the decision-maker is accepting a higher program/project risk, which will lead to the development of more challenging requirements and increased CRM burden regardless of which alternative is selected.

- **They present a risk of exceptionally poor performance** – This situation is the reverse of the situation above. In this case, even though the likelihood of not meeting the performance commitment is within the decision-makers' risk tolerance, the consequences may be severe, rendering such an alternative potentially unattractive.

Another uncertainty consideration, which is addressed below in the discussion of the iterative nature of deliberation, is whether or not a performance measure's uncertainty can be effectively reduced, and whether or not the reduction would make a difference to the decision. This issue is mentioned here because wide pdfs can lead to poor performance commitments relative to other alternatives, and it would be unfortunate to discard an alternative on this basis if additional analysis could be done to reduce uncertainty. Note that if two attractive alternatives present themselves and time and resources are available, it may be advantageous to proceed with, at least, partial prototyping (that is, prototyping of some of the critical components) of both to provide the necessary data for reducing key performance measure uncertainties such that a robust decision can be made.

3.3.2.4 Other Considerations

Depending on the decision situation and proposed alternatives, a variety of other risk-based, as well as non-risk-based, considerations may also be relevant. These include:

- **Sensitivity of the performance commitments to variations in risk tolerance** – Performance commitments are directly related to risk tolerance. Therefore, it is prudent for the deliberators to explore the effects of variations in the specified risk tolerances, to assure that the decision is robust to variations within a reasonable range of tolerances.

- **Risk disposition and handling considerations** – The risks that exist relative to performance commitments are ultimately caused by undesirable scenarios that are identified and analyzed in the risk analysis. Because of the scope of risk analysis for RIDM (i.e., the necessity to analyze a broad range of alternatives), risk retirement strategies may not be fully developed in the analysis. Deliberators' expertise is therefore brought to bear on the relative risk-retirement burdens that different alternatives present. For example, deliberators might feel more secure accepting a technology development risk that they feel they can influence, rather than a materials availability risk they are powerless to control.

- **Institutional considerations** – Different alternatives may have different impacts on various NASA and non-NASA organizations and institutions. For example, one alternative might serve to maintain a particular in-house expertise, while another alternative might help maintain a regional economy. These broad-ranging issues are not necessarily captured in the performance measures, and yet they are of import to one or more stakeholders. The deliberation forum is the appropriate venue for raising such issues for formal consideration as part of the RIDM process.

3.3.2.5 Deliberation Is Iterative

As illustrated in Figure 33, deliberation is an iterative process that focuses in on a set of contending alternatives for consideration by the decision-maker. Iteration during deliberation has both qualitative and quantitative aspects:

- **Qualitative** – A deliberator may have a particular issue or concern that he or she wishes to reach closure on. This might require several rounds of deliberation as, for example, various subject matter experts are called in to provide expertise for resolution.

- **Quantitative** – One or more performance measures might be uncertain enough to significantly overlap, thereby inhibiting the ability to make a robust decision. Moreover, large uncertainties will, in general, produce poor performance commitments, particularly when risk tolerance is low. Therefore, before a set of contending alternatives can be chosen, it is important that the deliberators are satisfied that particular uncertainties have been reduced to a level that is as low as reasonably achievable given the scope of the effort. It is expected that the risk analysis will be iterated, under the direction of the deliberators, to address their needs.

3.3.2.6 Communicating the Contending Alternatives to the Decision Maker

There comes a time in RIDM when the remaining alternatives all have positive attributes that make them attractive in some way and that make them all contenders. The next step is to find a way to clearly state for the decision-maker the advantages and disadvantages of each remaining alternative, especially how the alternatives address imposed constraints and satisfy stakeholder expectations. It is important that the process utilized by the deliberators affords him or her with ample opportunity to interact with the deliberators in order to fully understand the issues. This is particularly true if the decision-maker has delegated deliberation and downselection to a proxy. The information and interaction should present a clear, unbiased picture of the analysis results, findings, and recommendations. The more straightforward and clear the presentation, the easier it becomes to understand the differences among the alternatives.

Some of the same communication tools used in the TBfD can be used here as well, applied to the contending alternatives forwarded for the decision-maker's consideration. The imposed constraints risk matrix (Figure 30) summarizes what is among the most critical risk information. Additionally, information produced during deliberation should be summarized and forwarded to the decision-maker. This includes:

- **Risk tolerances and performance commitments** – The deliberators establish risk tolerances on the performance measures, for the purpose of generating performance commitments that can serve as the primary basis for comparison of alternatives. These tolerances and the resulting performance commitments are key pieces of information for the decision-maker. They strongly influence requirements development and the corresponding program/project risk that is to be accepted going forward. A notional performance commitment chart is shown in Figure 37.

Alternative \ Risk Tolerance	Performance Commitments				
	PM_1	PM_2	PM_3		PM_n
	10%	1%	30%	...	5%
1	100	0.007	0.01		0.4
2	500	0.0002	0.15		5
3	350	0.004	0.08		20
...				...	
m	400	0.0001	0.25		0.1

Performance Legend
Lower
Medium
Higher

Figure 37. Notional Performance Commitment Chart

- **Pros and cons of each contending alternative** – An itemized table of the pros and cons of each alternative is also recommended for the contending alternatives. This format has a long history of use, and is capable of expressing qualitative issues. It enables conflicting opinions to be documented and communicated to the decision-maker, so that he or she is aware of contentious issues and/or competing objectives among stakeholders.

- **Risk lists** – Each alternative will have different contributors to its performance commitment risks. Correspondingly, each contending alternative will have a risk list written for it that identifies the major scenarios that contribute to risk. Each scenario has the potential to impact multiple performance measures over multiple mission execution domains.

  Figure 38 presents a notional example of a RIDM risk list. Each row of Figure 38 represents a "risk," as the term is used in the CRM process. Each risk is articulated in terms of an existing *condition* that indicates a possibility of an undesired *departure* from the program/project baseline. The departure credibly impacts some program/project *asset*, resulting in a *consequence* that negatively affects the ability to meet one or more performance commitments. The magnitude of the impact is indicated in stoplight format (red/yellow/green) on a performance commitment basis, as well as on a holistic basis. The basis for determining the magnitude depends on the form of the risk assessment and the criteria established in the RMP, if one exists. For example, analyses that use detailed logic modeling might express risk contributions in terms of importance measures such as the Fussell-Vesely or Risk Reduction Worth (RRW) importance measures [27]. Less detailed analyses might use more qualitative criteria. Whatever method is used, consistency between the RIDM and CRM processes in this respect aids in the initialization of CRM for the selected alternative.

Alternative X – RIDM Risk List							
Risk #	Risk Statement	Performance Commitments					Total
		PM$_1$	PM$_2$	PM$_3$	...	PM$_n$	
1	Given A there is a possibility of B which impacts C and may lead to D	High	Medium	N/A		Low	High
2	Given E there is a possibility of F which impacts G and may lead to H	N/A	Low	Medium		N/A	Medium
3	Given I there is a possibility of J which impacts K and may lead to L	Medium	Medium	N/A		N/A	Medium
...					...		
m	Given W there is a possibility of X which impacts Y and may lead to Z	N/A	N/A	Low		Low	

Risk Legend

High
Medium
Low
N/A

Figure 38. Notional Risk List for Alternative X

In cases where performance measure pdfs have been generated using parametric modeling, the risk model might not explicitly enumerate the major factors contributing to performance risk, making the development of a risk list problematic. In such cases, there may be a need for iteration with the risk analysts to develop scenarios and identify such factors, so that the decision maker can be informed of the major specific engineering (and/or other) issues that go along with each alternative. This information is considered to be a vital input to a properly risk-informed decision making process.

Regardless of how well the risk information is summarized or condensed into charts or matrices, the decision-maker should also always be presented with the raw risk results, namely the performance measure pdfs, upon request. Only by having these fundamental analysis results can the decision-maker bring his or her full judgment to bear on the selection of an alternative. Band-aid charts, as shown in Figure 31, are appropriate communication tools for communicating this information to the decision-maker.

3.3.2.7 Alternative Selection Is Iterative

Just as risk analysis and deliberation iterate until the deliberators are satisfied that their issues and concerns have been satisfactorily addressed, alternative selection also iterates until the decision-maker is satisfied that the information at his or her disposal is sufficient for making a risk-informed decision. This is especially true in situations where the decision-maker has delegated much of the activity to others, and is exposed to the issues mainly through summary briefings of analyses and deliberations conducted beforehand. Iteration might consist of additional focused analyses, additional subject matter expert input, consideration of alternate risk tolerances (and associated performance commitments) for some performance measures, etc.

3.3.2.8 Selecting a Decision Alternative

Once the decision-maker has been presented with enough information for risk-informed decision making, he or she is ready to select a decision alternative for implementation. The decision itself consists of two main ingredients: the selection of the decision alternative and finalization of the performance commitments.

- **Selecting a decision alternative** – The RIDM process is concerned with assuring that decisions are risk-informed, and does not specify a particular process for selecting the decision alternative itself. Decision-makers are empowered to use their own methods for decision making. These may be qualitative or quantitative; they may be structured or unstructured; and they may involve solitary reflection or the use of advisory panels. Regardless of the method used for making the decision, the decision-maker formulates and documents the decision rationale in light of the risk analysis.

- **Finalizing the performance commitments** – In the requirements-based environment of the NASA program/project life cycle, decisions are essentially defined by the requirements they produce. Performance commitments capture the performance characteristics that the decision-maker expects from the implemented alternative, and also establish the initial risk that the decision-maker is accepting and calling on the CRM process to manage.

  As discussed in Section 3.3.1, performance commitments are produced by the deliberators as a result of establishing risk tolerances on the performance measures. This facilitates deliberation of alternatives in terms of point value estimates of performance that reflect the deliberators' risk attitudes. The decision-maker may choose to keep the risk tolerances and performance commitments established by the deliberators, or he/she may choose to modify them in accordance with his/her own risk tolerances. In situations where the decision-maker's risk tolerances differ significantly from those established by the deliberators, the decision-maker may ask for additional deliberation in light of the modified commitments. In turn, the deliberators may ask the risk analysts for a revised risk list that reflects the new situation.

3.3.2.9 Documenting the Decision Rationale

The final step in the RIDM process is for the decision-maker to document the rationale for the selected alternative in the RISR. In a NASA program/project context, the RISR is developed in accordance with the activity's RMP. Information on documenting the decision rationale can be found in Appendix D, Content Guide for the Risk-Informed Selection Report.

Planetary Science Mission Example:
Deliberate, Select an Alternative, and Document the Decision Rationale

Deliberation and selection of an alternative is done in light of the TBfD which, in addition to the results presented in the previous Planetary Science Mission Example boxes, also contains the imposed constraint risk matrix.

Planetary Science Mission Imposed Constraint Risk Matrix

Alternative	Time to Completion	Imposed Constraint Risk			Total*
		Project Cost	Data Volume	Planetary Contamination	
	Constraint (< 55 months)	Constraint (<$500M)	Constraint (> 6 months)	Constraint (< 0.1% prob.)	
1. Propulsive Insertion, Low-Fidelity Science Package	2.8%	22%	4.1%	1.1%	25%
2. Propulsive Insertion, High-Fidelity Science Package	2.4%	57%	6.4%	3.2%	62%
3. Aerocapture, Low-Fidelity Science Package	3.0%	9.7%	8.7%	5.5%	18%
4. Aerocapture, High-Fidelity Science Package	2.3%	47%	12%	12%	57%

*This is the probability of failing to meet one or more of the imposed constraints. Because the performance measures are correlated, the total probability is not necessarily the sum of the individual imposed constraint risk probabilities. For example, if *time to completion* is greater than 55 months, then *data volume* is zero.

The first objective of the deliberators is to see whether or not the set of alternatives can be pruned down to a smaller set of contending alternatives to present to the decision maker. The imposed constraint risk matrix shows that the risk of not meeting the $500M cost constraint is high for Alternatives 2 and 4 compared to the agreed-upon risk tolerance of 27%. Specifically, Alternatives 2 and 4 are infeasible given the combination of the cost constraint and the JCL policy, which specifies that the project be budgeted at the 70th percentile or greater. The 70th percentile cost estimates are $860M for Alternative 2 and $650M for Alternative 4. Thus, the deliberators prune these alternatives from contention.

Alternatives 1 and 3 are identified as the contending alternatives that are recommended to the decision maker for consideration.

Planetary Science Mission Example:
Deliberate, Select an Alternative, and Document the Decision Rationale (continued)

In choosing between Alternatives 1 and 3, the decision maker must weigh differing performance capabilities with respect to competing objectives. Alternative 1, which uses propulsive insertion, has the following pros and cons:

Alternative 1: Propulsive Insertion, Low-Fidelity Science Package

Pros:

- A relatively low risk of not meeting the 6 month *data volume* imposed constraint

- The ability to commit to a higher data volume, given the decision-maker's risk tolerance

- A low probability of planetary contamination that is well within the decision-maker's risk tolerance

Cons:

- Higher cost, due to the need for a medium size launch vehicle

Conversely, Alternative 3 has the following pros and cons:

Alternative 3: Aerocapture, Low-Fidelity Science Package

Pros:

- Use of aerocapture technology, which the decision-maker considers to be a technology that promises future returns in terms of reduced payload masses for missions that can exploit aerocapture and/or aerobraking opportunities

- Lower cost, due to the use of a small launch vehicle afforded by the lower payload mass

Cons:

- Higher risk of not meeting the 6 month *data volume* imposed constraint, due to the potential for aerocapture failure during insertion

- Lower data volume at the decision-maker's risk tolerance

- A higher probability of planetary contamination, though still within the decision-maker's risk tolerance

The decision maker sees the choice in terms of whether or not the Planetary Science Mission is appropriate to use as an aerocapture test bed. If Alternative 3 is chosen and the mission succeeds, then not only will it advance the technology of aerocapture, but it will save money for this and future missions. If it fails, then the only near-term opportunity to gather important data on Planet "X" will be lost. It is a difficult decision, particularly because Alternative 3's *data volume* imposed constraint is near the edge of the decision-maker's risk tolerance. The decision-maker confers with selected deliberators and stakeholders, chooses the alternative that he/she believes best balances the pros and cons of each contending alternative, and documents his/her decision rationale in the RISR.

4   THE CRM PROCESS

Once an alternative has been selected using the RIDM process and performance requirements have been developed for it as part of the technical requirements definition process [2], the risk associated with its implementation is managed using the CRM process. Because CRM takes place in the context of explicitly-stated performance requirements, the risk that the CRM process manages is the potential for performance shortfalls, which may be realized in the future, with respect to these requirements.

The CRM process manages risk by identifying specific issues that are of concern to one or more stakeholders, and which are perceived as presenting a risk to the achievement of one or more performance requirements. These risk-significant issues are referred to as *individual risks*, and collectively constitute the set of undesirable scenarios that put the achievement of the activity's performance requirements at risk. Each performance requirement then has an associated *performance risk* that is the aggregation of the risk impacts from the set of individual risks that threaten the requirement. Performance risk is quantified using risk analysis, using a scenario-based risk model that is informed from the risk model of the selected alternative developed during the RIDM process.[14] This risk model is augmented and refined as needed throughout implementation as individual risks are identified and incorporated into the model. In this way, the risk model is maintained as a quantitative tool for understanding the cumulative impacts that individual risks have on the performance requirements.

The individual risks considered in this version of the RM Handbook are primarily program/project risks, although some attention is also given to institutional risks. Institutional risks, as defined in NPR 8000.4A, are: "risks to infrastructure, information technology, resources, personnel, assets, processes, occupational safety, environmental management, or security that affect capabilities and resources necessary for mission success, including institutional flexibility to respond to changing mission needs and compliance with external requirements (e.g., Environmental Protection Agency or Occupational Safety and Health Administration regulations)." Unlike program/project risks, institutional risks do not start from a systems engineering process. Because they are substantially different in character, institutional risks will be treated in more detail in a later addendum to this handbook.

As in RIDM, risk analysis using an integrated risk model (as practicable per a graded approach) is central to CRM. This is because:

- The risk to a given performance requirement (i.e., the requirement's performance risk) may be due to the cumulative effects of a number of individual risks, rather than being due to a single individual risk. An integrated risk model is performance-risk-centric in that it identifies significant performance risk even when the sources of that risk are distributed.

---

[14] Refer to Section 2.2.1 for a discussion of how some of the parameter based risk models developed in RIDM may need to be converted to scenario based risk models during the initialization of CRM.

- Dependencies among individual risks can be uncovered using an integrated risk model. The effects of a given individual risk may be exacerbated or inhibited by the presence of a different individual risk. Integrated risk modeling allows these dependencies to be characterized and reflected in the resulting performance risk.

- Integrated risk modeling supports the identification of *risk drivers*. Risk drivers are those elements found within the aggregate performance risk models that contribute most to the performance risks because of uncertainties in their characterization. As discussed in the blue box below, they can be events in a risk scenario or performance parameters that affect the probabilities of those events. An event or a parameter may be a risk driver due to its role in a significant individual risk, or it may be a risk driver due to its combined influence in a number of individual risks.

Risk Drivers

A risk driver is a significant source of performance risk. Operationally, a risk driver can be a single performance parameter, a single event, a set of performance parameters collectively, or a set of events collectively that, when varied over their range of uncertainty, causes the performance risk to change from tolerable to intolerable (or marginal). As such, risk drivers focus risk management attention on those potentially controllable situations that present the greatest opportunity for risk reduction. Often, risk drivers affect more than one individual risk and cut across more than one organizational unit.

Risk drivers can be communicated at various levels of the risk model. At the lowest level of the model they can be expressed as the possibility that the performance parameters will be (or are) outside the range of tolerable performance risk. Alternatively, inasmuch as the risk driving performance parameters affect the outcome of pivotal events in the risk model, the risk drivers can be expressed as the possibility that certain undesirable pivotal events will occur.

Risk drivers are identified during the *Analyze* step of CRM and are used during the CRM *Plan* step to devise effective risk response options. The nature of the risk response options to be considered for a given risk driver depends upon the nature of the uncertainty that makes it a driver. If the uncertainty is primarily aleatory (i.e., originates from intrinsic variability in the system and/or its environment), risk reduction is typically accomplished via mitigation (i.e., positive steps to reduce or eliminate the undesired variability). If the uncertainty is primarily epistemic (i.e., originates from lack of knowledge), then the response usually starts with research to better determine the facts of the matter, and proceeds to mitigation only if the risk after completion of the research is still intolerable.

- Integrated risk modeling supports performance-risk-centric risk responses. The CRM Plan step entails the generation and risk analysis of candidate risk response alternatives, i.e., the proposed strategies for managing performance risk so that it is within tolerable levels. Risk analysis shows how each risk response alternative impacts the entire set of modeled performance measures, for a full characterization of their various impacts on performance risk.

- Risk models and risk analysis results can be coordinated among the units of the NASA organizational hierarchy, so that each unit's risk analysis is integrated with the analyses of subordinate units that directly affect the unit's performance risk, as well as with other units that share common sources of risk.

The tolerability of each requirement's performance risk is set at the outset by a decision maker within the organizational unit that is responsible for that risk and, depending on the viewpoint of the decision maker, the tolerability may be specified as a function of time. At the outset of implementation, the risk tolerances generally reflect those established by the RIDM process to the extent that the performance requirements reflect the RIDM performance commitments. However, as implementation proceeds, the decision makers may stipulate that some of the performance risks should be burned down according to established risk burn-down schedules, so that as the verification target date approaches, the confidence that the performance requirement will be met becomes increasingly high. The decision makers may also stipulate, however, that other performance risks should remain constant throughout the project, or even increase over time via a prescribed relaxation schedule. Some of the principles behind risk burn-down or relaxation will be discussed further in Section 4.1.3.

As shown in Figure 39, the CRM process consists of the five cyclical steps, *Identify*, *Analyze*, *Plan*, *Track*, and *Control*, supported by comprehensive *Communicate* and *Document* functions. In practice, these steps operate in parallel, such that at any given time there may be individual risks being reported into the risk database; other individual risks being incorporated into the risk model; risk response plans being developed to reduce performance risk to tolerable levels; and implemented risk responses being tracked and controlled as needed to ensure their desired effects.

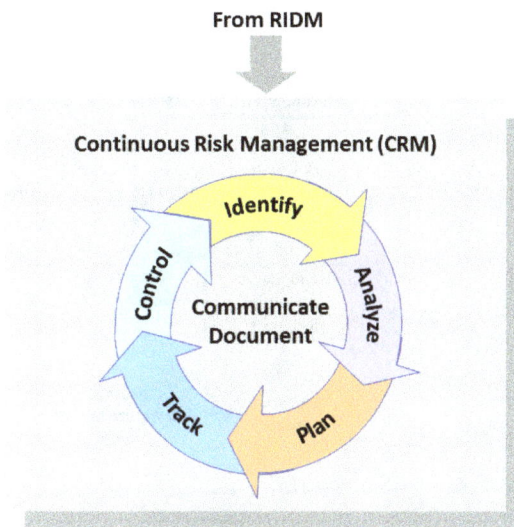

Figure 39. The CRM Process

Moreover, the activities of each step, and the interfaces between them, depend on the nature of the risk issues being addressed, such that different issues can follow somewhat different paths through the process. Figure 40 illustrates the situation. Beginning with the *Identify* step, individual risks are generated, due either to their prior identification during RIDM or to their identification during implementation. These individual risks are given a "quick look" analysis to determine their urgency. Urgent risks are forwarded immediately to the *Plan* step so that a timely risk response can be implemented, while at the same time being analyzed further via integration into the risk model. Risks that are not urgent are analyzed in detail prior to planning, to ensure that planning is appropriately risk-informed using an analytical basis that is sufficient to support robust selection of effective risk responses. In either case, the risk model is updated with the selected risk response, and the risk drivers are tracked and controlled as necessary to keep performance risk within tolerable levels. If, however, there are inadequate resources at the current level to implement an effective tactical or strategic response, the risk decision is elevated to the next higher organizational unit. Additionally, if an effective strategic response cannot be implemented at any organizational level, the performance requirements are reevaluated by reopening the RIDM and SE processes.

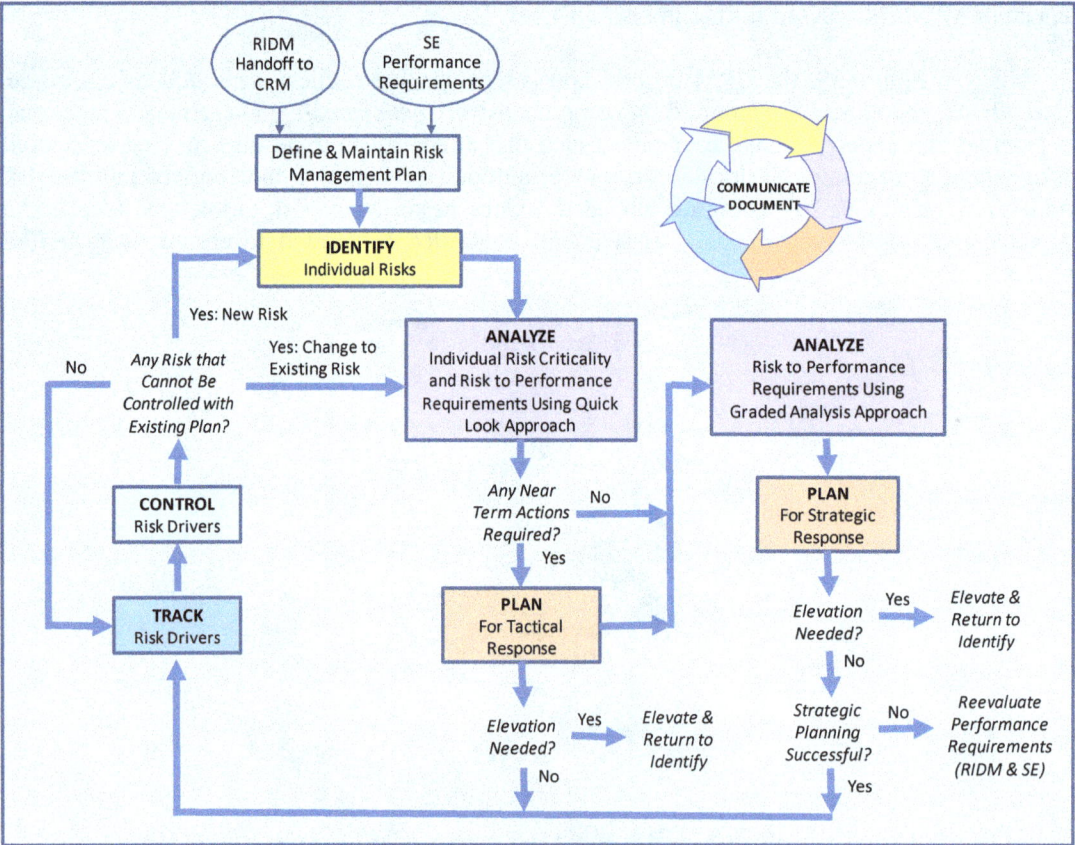

Figure 40. CRM Process Flow Diagram

The above discussion and figure distinguishes between two different dimensions of CRM: a tactical dimension and a strategic dimension. The distinction is further discussed in the blue box below.

Tactical and Strategic Dimensions of CRM

NPR 8000.4A states that "risk is the potential for performance shortfalls, which may be realized in the future, with respect to achieving explicitly established and stated performance requirements." In this Handbook, the risk to achieving performance requirements is termed *performance risk*. Because the risk to a given performance requirement may have many sources, all of which interact to produce the aggregate performance risk, effective management of performance risk requires integration of these risk sources into a quantitative risk analysis (i.e., one that is quantitative to the extent that is practical and needed), in order to properly understand them and assess the effectiveness of proposed risk management responses to them. However, during implementation, conditions can arise that indicate individual sources of significant risk, to which only a limited timeframe exists for effective risk management response. Such situations must be responded to quickly, on a shorter time scale than that required for quantitative risk analysis, so that the window of opportunity for response is not lost.

Consequently, the CRM process involves both a "tactical" dimension and a "strategic" dimension:

- The *"tactical"* dimension of CRM supports rapid responses to newly identified sources of potentially significant risk. It conducts risk analysis and planning primarily in the context of the *individual risks* that are reported into the risk database.

- The *"strategic"* dimension of CRM supports a comprehensive understanding of the sources of *performance risk*. It conducts risk analysis and planning primarily in the context of a quantitative risk model that integrates all known sources of significant performance risk.

4.1 Initializing the CRM Process

After an alternative has been selected for implementation using the RIDM process, and performance requirements have been defined for it by the systems engineering process, the CRM process is initiated to provide a framework for ongoing risk management in a manner that is appropriately standardized across the affected organizational units.

4.1.1 Development of Risk Management Plan

Even though it is not explicitly included in the traditional representation of the CRM process, an upfront phase of planning is necessary to assure the development of a robust risk management process. Once it is developed, the risk management process is formally documented in a project RMP (or where appropriate, a program RMP). In addition to detailing how each risk management process step will be carried out, the RMP should also serve as the means to identify and define the key coordination and technical provisions that are to be implemented in the course of the risk management programmatic activities.

The key elements of the risk management framework that are developed and captured in the RMP include:

- Identification of stakeholders, such as Risk Review Boards, to participate in deliberations regarding the response to risks

- Establishment of risk tolerance criteria, thresholds, and elevation protocols (the specific conditions under which a risk management decision must be elevated through management to the next higher level)

- Establishment of a risk tolerance schedule for each performance risk (to be developed further in Section 4.1.3)

- For each performance requirement, documentation, or indication by reference, of whether its associated risks (including the aggregate risk) are to be assessed quantitatively or qualitatively and provides a rationale for cases where it is only feasible to assess the risk qualitatively

- Establishment of risk communication protocols between management levels, including the frequency and content of reporting, as well as identification of entities that will receive risk tracking data from the unit's risk management activity

- Delineation of the processes for coordination of risk management activities and sharing of risk information with other affected organizational units.

Practical guidance is provided in Appendix F on aspects of the RMP involving staffing resources, plan updates, training, and interaction of the project and contractor processes.

4.1.2 Inputs to CRM

4.1.2.1 Inputs from the RIDM Process

Many of the products of the RIDM process carry over to CRM. These products include:

- The risk analysis (Section 3.2) of the selected alternative – The risk analysis that was developed during RIDM for the selected alternative is maintained throughout the CRM process. It provides the core risk analysis capability for assessing performance risk and identifying risk drivers.

- Risk tolerances and associated performance commitments – The risk tolerances on the performance measures represent the initial performance risk that the decision maker implicitly accepted as part of the development of risk-normalized performance commitments (Section 3.3.1). The performance commitments represent the levels of performance that can be achieved by each performance measure at its associated risk tolerance.

- The risk list – The risk list generated during RIDM identifies the major scenarios contributing to the selected alternative's performance risk, with the caveat that during RIDM, performance risk is defined in terms of performance commitments rather than performance requirements. Performance requirements are developed by systems engineering after the selection of an alternative for implementation. The expectation is that the performance requirements will have been derived in light of the performance commitments, thus assuring that they are achievable within the risk tolerance of the decision maker.

- The Risk-Informed Selection Report (RISR) – The RISR documents the RIDM process, including the decision rationale, as discussed in Section 3.3.2.9 and Appendix D.

The risk analysis will have addressed all of the performance measures for the selected alternative, but only those performance measures considered discriminators among the alternatives are likely to have been analyzed in detail. Moreover, because the initial risk list is based on the RIDM risk analysis, it is likely to contain only the major, top-level, initially evident risks and may therefore be incomplete, especially with respect to the non-discriminator performance measures.

As soon as feasible, the CRM process will need to complete the RIDM risk analysis for the non-discriminator performance measures and expand and update the initial risk list to include any new risks from the completed risk analysis. After the establishment of performance requirements by Systems Engineering, the risk analysis will have to be updated to reflect the performance requirements and resulting risks relative to them. At this time, there will also be a transition from parameter based modeling to scenario based modeling wherever feasible, as discussed in Section 2.2.1.

The risk models developed during RIDM are essentially self-contained at every level of the organizational hierarchy. Indeed, because RIDM is often conducted during Formulation, RIDM risk analyses can be done before the organizational hierarchy is established. This is not true of CRM risk models, because CRM is conducted at every level of the hierarchy. Therefore, CRM risk models rely on the upward flow of risk information from the models at lower levels of the organizational hierarchy, as well as on the consistency of models and assumptions among all units whose results are ultimately aggregated into performance risk at some shared higher level. Consequently, model sharing and data reporting protocols must be established as needed to support the distributed nature of risk management within the hierarchical structure of NASA. Section 4.3.3.1.2 addresses the integration of risk models between levels in the NASA hierarchy, and Section 4.7.1 addresses the communication protocols necessary to ensure consistency and availability of data among units that are subordinate to a common set of objectives

4.1.2.2 Inputs from Systems Engineering

The output from Systems Engineering that constitutes input to the CRM process consists of the performance requirements for the selected alternative. These performance requirements are the result of a negotiated requirements decomposition and allocation process that flows downward through the NASA hierarchy, as discussed in the NASA Systems Engineering Handbook [2]. It is expected that the requirements will be risk informed, based on RIDM's performance measure pdfs, imposed constraints, risk tolerance levels and associated performance commitments, the risk analysis of the selected alternative, and any other information deemed pertinent by the Systems Engineering decision-maker. However, the performance requirements, including the parameters upon which requirements are levied, will not necessarily be identical to the performance measures and performance commitments of RIDM. The systems engineering process is not constrained to base its requirements on the RIDM analysis, only to consider the risk information it produces in its deliberations.

4.1.3   Risk Tolerance Targets at Projected Milestones

As discussed in Section 4, the initial risk levels for each performance measure establish initial *risk tolerance levels*[15] for the achievement of performance requirements. Often, the expectation of improvement with time leads to a tightening of tolerance levels according to a risk burn-down schedule at key program/project milestones, which is combined with the objective of meeting the requirements per an associated *verification standard*[16]. In other words, as the program/project evolves over time, mitigations are implemented; and as risk concerns are retired and the state of knowledge about the performance measures improves, uncertainty should decrease, with an attendant lowering of residual risk (see Figure 41). The decrease may not be linear, as new risks may emerge during the project requiring new mitigations to be instituted.

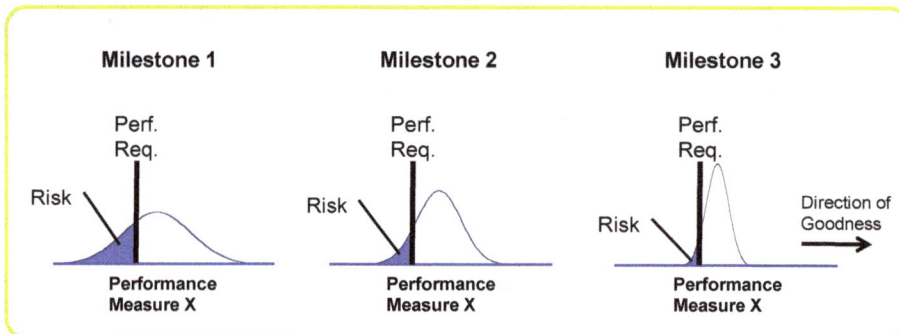

Figure 41. Decreasing Uncertainty and Risk over Time

As an example, the risk of not meeting the required launch date may be relatively high at the beginning of the project if the project requires the development of new technology. The RMP will specify that this risk should diminish toward zero by the time the required launch date occurs.

In other cases, the decision maker may believe that the tolerance levels should remain unchanged or decrease with time. For example, he/she may believe that vehicle mass or project cost requirements should be controlled by specifying margins that are initially set very high in order to compensate for high uncertainty, and that are subsequently relaxed as the launch date approaches. Reducing a performance margin is equivalent to relaxing a risk.

Because the performance requirements are risk-informed but developed outside of the RIDM process, it is possible that their values may differ significantly from the performance commitments. In some cases the performance requirements may be more stringent than the performance commitments and may require higher risk tolerance levels compared to the

---

[15] Risk tolerance levels are levels of risk considered to be tolerable by a decision maker at a given point in time.
[16] Verification standards are the standards used to verify that the performance requirements have been met.

performance commitments. *Note: In such cases it may be prudent for program/project management to negotiate new performance requirements with System Engineering.*

Since for the above reasons, the performance requirement values may differ from the corresponding performance commitment values of the selected alternative, the RIDM analysis process needs to check how any such difference translates into initial program risk tolerance levels. This is done by comparing the performance requirement values with the performance measure pdfs that were initially used to establish the performance commitments. The initial risk tolerance levels corresponding to the established performance requirements are transmitted to CRM, together with a schedule for their burn-down or relaxation.

Figure 42 notionally shows a set of risk burn-down schedules. Each performance requirement has its own burn-down schedule that is established by the decision maker within each organizational unit and included in the RMP as discussed above. The profile is based on the initial assessed performance risk and the timeline during which that risk is expected to be retired as implementation proceeds and issues that threaten their achievement are managed. As such, the details of the burn-down schedule depend on the details of the project plan. As intermediate milestones in the plan are accomplished, events associated with the non-achievement of the milestones no longer threaten the project plan and the remaining performance risk drops. In this way, the nominal success path through the project plan defines the tolerable risk at every point in time during implementation.

The schedules in Figure 42 show three notional grades of risk tolerance: *tolerable*, *marginal*, and *intolerable*.[17] Ideally, every performance risk begins in the *tolerable* range. However, to the extent that the performance risks differ from the corresponding RIDM performance commitments, or that additional sources of risk are identified during CRM initialization, there may be performance risks that begin in the *marginal* or even the *intolerable* range. As time passes, levels of performance risk that are initially *tolerable* may become *marginal* and then *intolerable*, because they no longer reflect the residual risk associated with the project plan's burn-down profile.

4.1.3.1 Alternate Performance Margin Targets at Projected Milestones

As discussed above, the decision maker may, in certain situations, prefer to specify tolerability in terms of performance margins instead of risks. This would generally be the case when the decision maker believes that the tolerability for certain risk requirements should be relaxed as the project gains maturity. Allowing the decision maker this flexibility also benefits the project because the design and system engineering organizations are familiar with the practice of controlling risks by controlling margins.

---

[17] The concept of marginal risk is notional for purposes of this handbook, and is intended to allow for the idea that the boundary between tolerable and intolerable may have a "fuzzy" aspect.

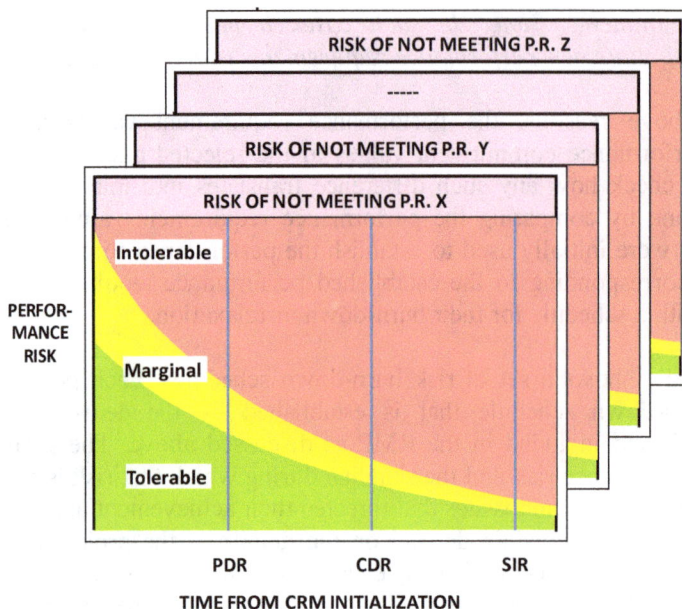

Figure 42. Notional Risk Burn-Down Schedules for Several Performance Requirements

Margins apply to performance measures that are continuous variables (e.g., cost, time to launch, payload mass). A definition of performance margin is presented in the blue box below.

Performance Margins

A performance margin is defined as the distance from the achieved value of a performance measure at any point in time to a decision boundary for that measure. The decision boundary may be the performance requirement, or it may be some other value related to the performance requirement. For example, "mass margin" for a payload may be defined as the difference between the maximum value allowed by a performance requirement and a calculated value of the payload mass using some set of assumptions. The decision boundary in this case is the performance requirement. Sometimes, however, the maximum allowed value is taken to be the performance requirement minus a contingency percentage of that value (e.g., 10%). The contingency is intended to allow for mass growth that is expected to occur, based on prior experience, from the time the system is initially designed to the time it is actually fielded. In that case the decision boundary is 90% of the performance requirement.

In the context of risk, just as each performance measure is treated as a random variable having an uncertainty distribution (e.g., Figure 41), so is each performance margin. That is because the margin is simply the performance measure minus a constant value (the decision boundary).

A typical statement on risk tolerability could be phrased as follows:

"The probability that performance requirement X will not be met at a specified decision point during the project shall be less than 10%."

Correspondingly, a typical statement on margin tolerability could be phrased as follows:

"The margin for performance measure x at a specified decision point during the project shall be greater than y at a 90% confidence level."

It should be recognized that the two statements are identical when $y = 0$.

The term "confidence level" has a statistical origin, but for purposes of this discussion it should be considered to be the complement of the uncertainty level. That is, a 90% confidence level is equivalent to 10% uncertainty as to whether the cited margin can be achieved.

Figure 43 conceptually illustrates one means for stipulating burn-down schedules for performance margins. As was discussed relative to Figure 42 for performance risks, each performance margin has its own burn-down schedule that is established by the decision maker within each organizational unit and included in the RMP. The principal difference between this figure and Figure 42 is that while the margin is allowed to decrease as the project becomes more mature, the confidence level associated with the margin is expected to increase. The decision maker specifies both the margin thresholds and the confidence level as a function of time.

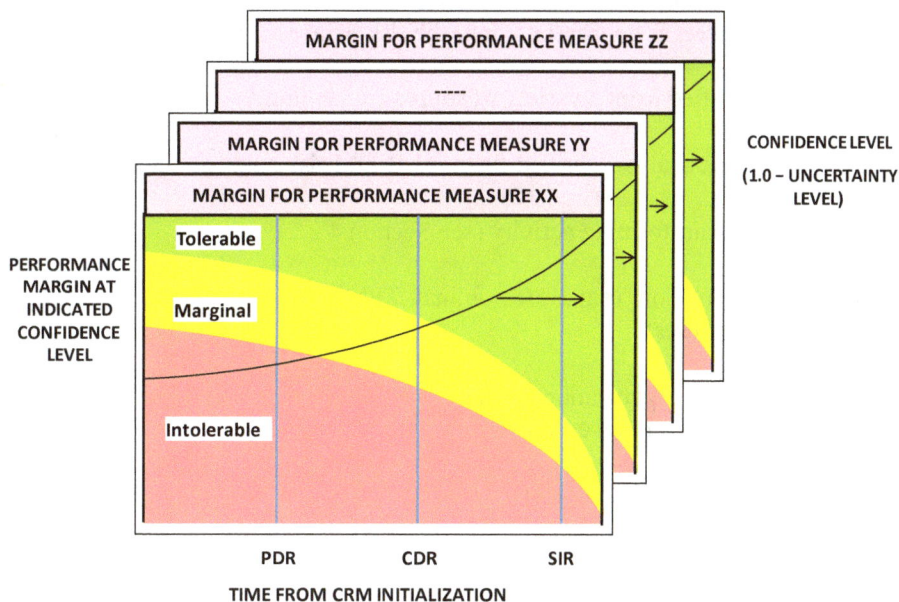

Figure 43. Notional Margin Burn-Down (Risk Relaxation) Schedules for Several Performance Margins

4.1.4 Developing Initial Risk Taxonomies

Taxonomies can be used to help identify risks that otherwise might be missed, to help categorize risk drivers, to help define effective mitigation alternatives, and to assist the project in properly allocating resources amongst organizational units. A *taxonomy* is a tree structure of classifications that begins with a single, all-encompassing classification at the root of the tree, and partitions this classification into a number of sub-classifications at the nodes below the root. This process is repeated iteratively at each of the nodes, proceeding from the general to the specific, until a desired level of category specificity is reached. Guidelines for taxonomy development will be provided in Section 4.2.1.3.

4.2 The CRM Identify Step

The objective of the CRM *Identify* step is to identify issues that potentially threaten the achievement of requirements, and incorporate them into the risk management process so that they can be addressed in a timely manner before they become problems. The Identify step is the input of the risk management process; it captures concerns, structures them as *individual risks*, and assigns them to the organizational units of the NASA hierarchy that are initially considered best suited to manage them.

Individual risks are ideally identifiable by all project personnel. Access to the risk database should be correspondingly liberal, to enable all persons to bring risk management attention to concerns they may have about issues that threaten project success. A corollary of this open-access philosophy is that the inputs to the risk database must be vetted by risk management personnel, who work with the identifying personnel to understand their concerns and articulate them in an effective manner, as appropriate, into the risk database.

Individual risks are identified from a variety of sources, including:

- The baselined risk analysis derived from the RIDM risk analysis activity

- The taxonomy brainstorming activity (see Section 4.2.2)

- Project implementation (e.g., off-nominal TPM data, off-nominal tracking data for existing risk responses)

- Inputs from other organizational units (e.g., assignment of newly identified issues, elevation of issues from lower units).

Figure 44 illustrates the CRM Identify step.

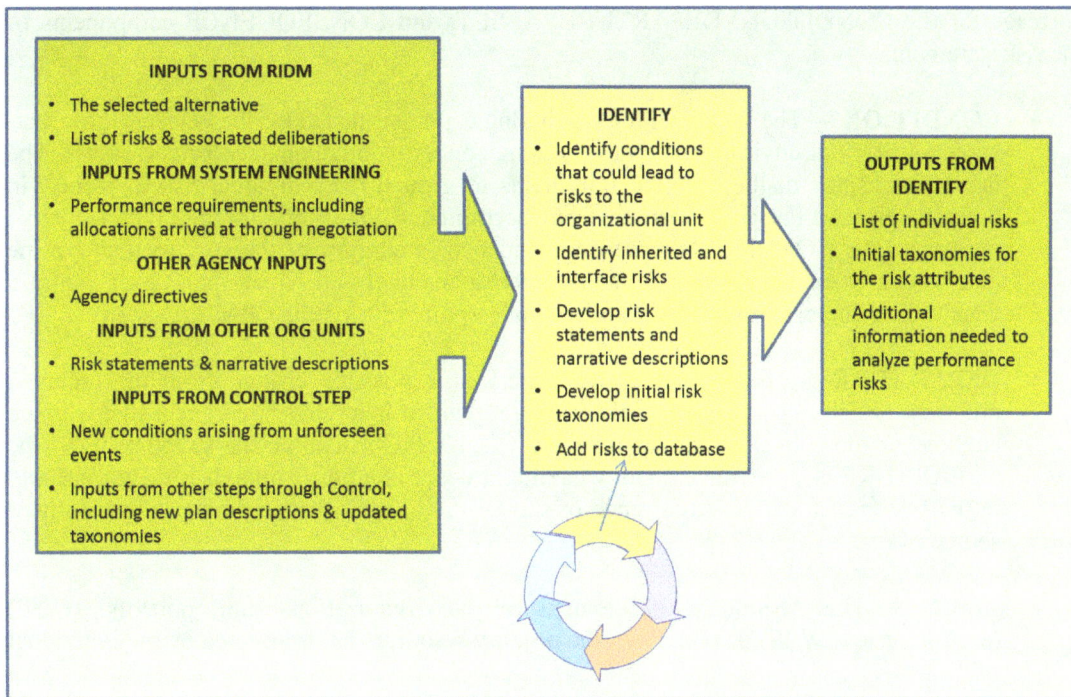

Figure 44. The CRM Identify Step

4.2.1 The Structure of an Individual Risk

Individual risks consist of two distinct parts: the ***risk statement*** and the ***narrative description***. The risk statement is a concise, structured description of the issue underlying the risk identifier's concerns. The narrative description captures the context of the risk statement, including circumstances, events, and interrelationships within the project that relate to it.

The risk statement captures only one part of the risk definition presented in Section 1.3, i.e., the scenario. The scenario is the part that is of interest to the CRM *Identify* step. The other two parts in the Section 1.3 risk definition, i.e., likelihood and consequence, concern the quantification of the magnitude of the risk. These are discussed in Section 4.3, the CRM *Analyze* step.

4.2.1.1 The Risk Statement

The risk statement has the following format:

"Given that [CONDITION], there is a possibility of [DEPARTURE] adversely impacting [ASSET], thereby leading to [CONSEQUENCE]."

It is the job of the risk identifier, working as needed with risk management personnel, to develop verbiage for the CONDITION, DEPARTURE, ASSET, and CONSEQUENCE components of the risk statement.

- **CONDITION** – The CONDITION is a single phrase that describes current key fact-based situation or environment that is causing concern, doubt, anxiety, or uneasiness. The fact-based aspect of the CONDITION helps to ground the individual risk in reality, in order to prevent the risk database from becoming a repository for purely speculative concerns. The CONDITION represents evidence in support of the concern that can be independently evaluated by risk management personnel and which may be of value in determining an appropriate risk management response during the CRM *Plan* step.

- **DEPARTURE** – The DEPARTURE describes a possible change from the (agency, program, project, or activity) baseline project plan. It is an undesired event that is made credible or more likely as a result of the CONDITION. Unlike the CONDITION, the DEPARTURE is a statement about what might occur at a future time. It is the uncertainty in the occurrence or non-occurrence of the DEPARTURE that is the initially identified source of risk.

- **ASSET** – The ASSET is an element of the organizational unit portfolio (OUP) (analogous to a WBS). It represents the primary resource that is affected by the individual risk.

- **CONSEQUENCE** – The CONSEQUENCE is a single phrase that describes the foreseeable, credible negative impact(s) on the organizational unit's ability to meet its performance requirements. It should describe the impact(s) in terms of failure to meet requirements that can be measured, described, and characterized.

The structure of the risk statement is somewhat longer than in existing paradigms, where the ASSET and CONSEQUENCE are combined into one element (CONSEQUENCE). In some formats, the DEPARTURE is also combined within the CONSEQUENCE element. In the latter case, the risk statement is simplified to the following: "Given that [CONDITION], it is possible that [CONSQUENCE]." This handbook prefers the longer format for two reasons:

- Separation of ASSET from CONSEQUENCE makes it easier to identify the organizational unit that is responsible for the risk, and also facilitates the construction of risk taxonomies (to be discussed in Section 4.2.1.3).

- Separation of DEPARTURE from CONSEQUENCE makes it necessary to treat the individual risk as a scenario, where the DEPARTURE is an event that must occur for the CONSEQUENCE to result, and the CONSEQUENCE is the inability to meet a performance requirement.

It is of fundamental importance to the CRM process that risk statements be crafted without regard to potential mitigations or other risk responses that may suggest themselves to the risk identifier. The risk statement should not presume anything that is not in the current baseline project plan, other than the CONDITION, which has its basis in fact. In particular, the CONSEQUENCE should presume that no risk response has been implemented that would shift the consequence from, say, a potential over-mass condition to a cost overrun and/or schedule slippage resulting from an anticipated redesign. It is recognized that the resulting risk statements can be considered artificial in that the issue might not be allowed to persist without a risk management response of any kind, but the point of the *Identify* step is specifically to capture the concern, not to presume the manner in which it will be addressed. This is not to say that the risk identifier should be silent on the topic of potential risk responses. On the contrary, such input is strongly encouraged, but should be included in the narrative description section of the individual risk, as will be discussed in Section 4.2.1.4.

An example of a risk statement for the planetary science mission is presented below. Other examples to be used in the CRM development for this handbook may be found in Section 4.2.1.3 and in Appendix G.

Risk Statement Example for Planetary Science Mission

Given that [CONDITION: the state of knowledge of Planet X's atmosphere is limited; the fact that it is difficult to ascertain more information about Planet X's atmosphere from Earth; and the fact that the spacecraft contains radioactive material], there is a possibility of [DEPARTURE: unanticipated atmospheric characteristics during the aerocapture maneuver at Planet X leading to a less-than-optimal trajectory] adversely impacting [ASSET: the spacecraft], thereby resulting in [CONSEQUENCE: spacecraft breakup and radioactive contamination of Planet X]

4.2.1.2 Validating an Individual Risk

The following eight questions can be used to guide the writing of an individual risk to ensure that it is valid. If the answer to any of the questions is "no" or "unknown," the risk should not be considered valid and the author may wish to go back and modify it appropriately (possibly with the help of the appropriate risk management personnel) or abandon the effort.

1. Does the individual risk (i.e., risk statement and narrative description) adequately communicate the possible sequence of events leading from the CONDITION through the DEPARTURE to the ASSET and the CONSEQUENCE?

2. Is the individual risk based on relevant documentation or individual/group knowledge?

3. Does the individual risk involve a change from the program/project/activity baseline plan for which an adequate contingency plan does not exist?

   Note: If it involves an existing contingency plan that is believed to be inadequate, the failure of that contingency plan should be addressed in the DEPARTURE portion of the risk statement.

4. Is the CONDITION factually true and supported by objective evidence?

5. Is the DEPARTURE credible (possible)?

6. Does the individual risk impact at least one agency/program/project/activity requirement that can be objectively measured, described, and characterized?

7. Is the CONSEQUENCE written without regard to potential mitigations?

8. Is the individual risk actionable (i.e, can something be done to prevent, or reduce the likelihood of the DEPARTURE and/or severity of the CONSEQUENCE)?

Determination of whether a risk is actionable (Item 8 above) is based on current assumptions about funding and other programmatic constraints. If fundamental changes are imposed on these current assumptions (e.g., major program cuts or cancellations), then it is necessary to go back into RIDM to determine whether the identification of alternatives and the selection from among them need to be changed.

After completion by the author and entry into the risk database, the individual risk is reviewed by risk management personnel using the same eight validity test questions. If the answer to any is "no" or "unknown," the individual risk is not considered valid and the author should be queried for his/her intent so that it can be either modified or rejected, with rationale. Figure 45 illustrates the process for entering and validating an individual risk.

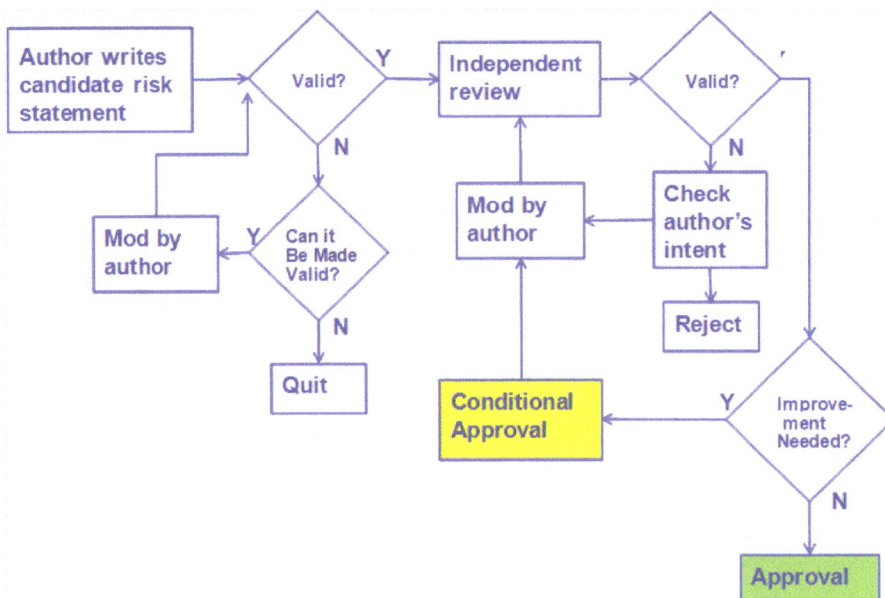

Figure 45. Generating and Validating an Individual Risk

The Measure of a Consequence

The *measure* of a consequence can be expressed in two different forms:

- The first form emphasizes the one-to-one relationship between consequence and performance risk: The measure is the conditional likelihood that the performance requirement will not be achieved, given the occurrence of the departure.

- The second form emphasizes the relationship between the consequence and the performance margin. The measure is the margin between the achievable value of a performance measure and the required value, given a specified confidence level.

The difference between the two consequence forms is illustrated schematically in the example below. In the charts below, the uncertainty distribution for the performance measure is shown in the form of a complementary cumulative distribution function (CCDF), which is obtained by integrating the probability density function (PDF) from right to left. A given value on the ordinate scale is equal to the probability that the actual value of the performance measure is greater than the corresponding value on the abscissa scale. The performance requirement of interest is that the launch date occur within a prescribed window. The measure of consequence in the left chart is the conditional likelihood of not being able to launch within the window, which according to the chart has a value of about 0.0004 (or about 0.04%). The measure in the right chart is the available margin at a 90% confidence level, which according to the chart is about 4 months (i.e., 6 months minus 2 months). In this Handbook, the first form will be used as the measure of consequence unless otherwise specified.[18]

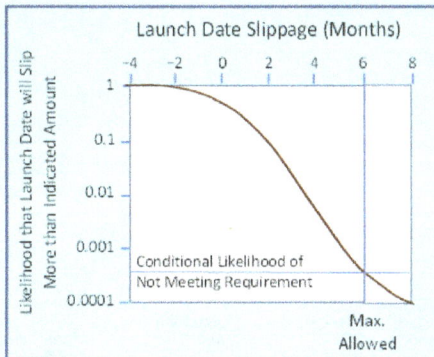

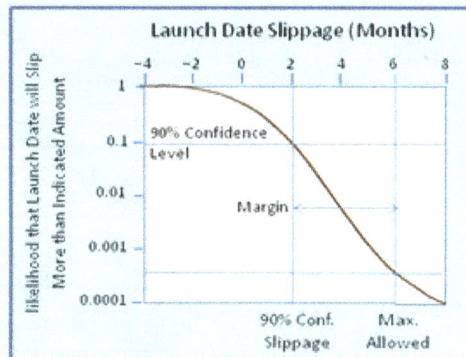

---

[18] Relative to the blue box above, in order to account for the fact that there may be interactions or dependencies between different individual risks (for example, they may share a common departure event), the *magnitude* of a consequence has to be defined differently from the *measure* of a consequence. The magnitude of a consequence is the difference between the measure of consequence obtained when all the individual risks are included in the performance risk model and the measure obtained when the new individual risk is excluded. For example, if the effect of the new individual risk is to increase the likelihood of not meeting a requirement from 0.04% to 0.05%, given the departure event for the new risk occurs, the magnitude of its consequence is a 0.01% increase in the likelihood of not meeting the requirement. Similarly, the magnitude of its consequence expressed in terms of margins would be a one month effect if the new individual risk caused the launch date margin to decrease from 4 months to 3 months at the specified confidence level, given occurrence of the departure event. Further discussion on the magnitude of a consequence will appear in Section 4.3.2.1.

4.2.1.3 Taxonomic Categorization of Individual Risks

The CRM process involves three distinct taxonomies, a condition/departure taxonomy, an asset taxonomy, and a consequence taxonomy. To the extent that different risk issues have elements in common, these commonalities can be identified by classifying each element according to the appropriate taxonomy. It may then be possible, when deciding on an appropriate risk management response to the identified risks, to craft individual responses that simultaneously address all or most of the elements within a given taxonomic category.

Figure 46 and Figure 47 show example condition/departure and asset taxonomies, respectively. Such taxonomies are established at CRM initiation and should be maintained and updated over the course of the project. In order for taxonomies to be useful in identifying cross-cutting risks, taxonomies of a given type should be uniform across the scope over which cross-cutting risks are a concern, e.g., the program or project. The consequence taxonomy can be derived directly from the requirements tree for the program/project, since, by definition, the consequence of interest with respect to a risk issue is its effect on the likelihood of meeting the performance requirements.

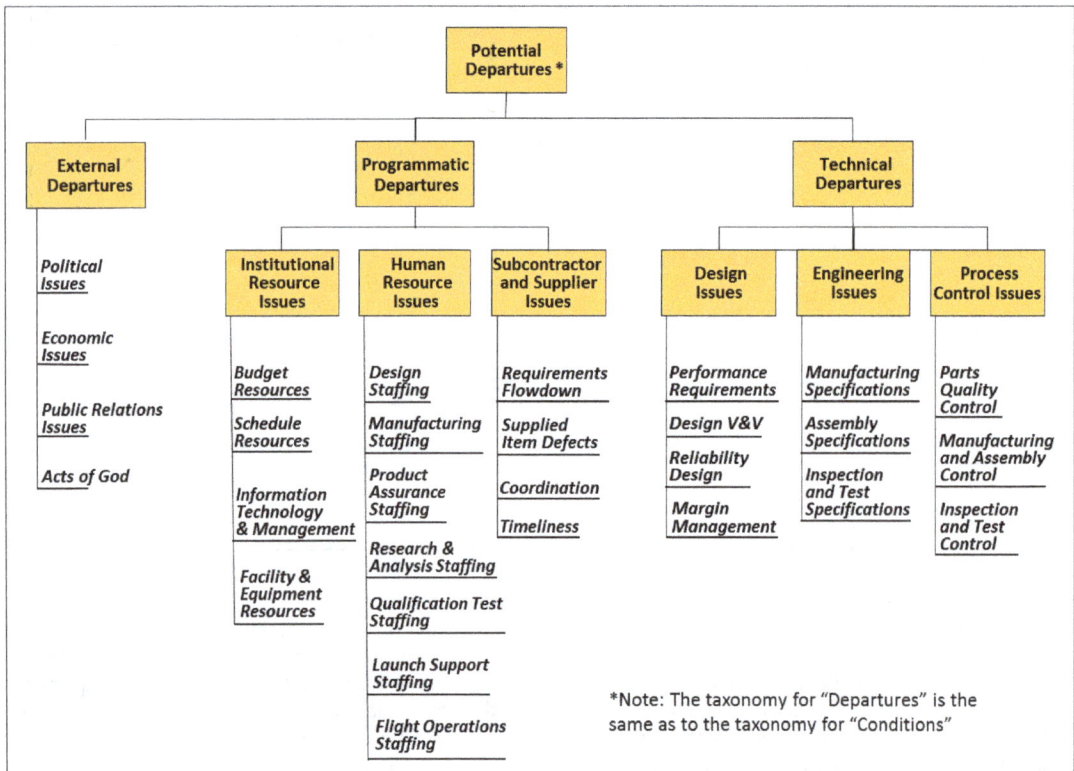

Figure 46. Example Condition/Departure Taxonomy

Potentially Affected Assets

- **External Assets**
  - Public Lives
  - Public Land
  - Tangible Property
  - Other Planets
- **Program-Wide Assets**
  - **Personnel**
    - Crew
    - Other Personnel
  - **Facilities and Equipment**
    - ...
    - ...
  - **Subcontractors and Suppliers**
    - ...
    - ...
- **Mission-Specific Assets**
  - **Space Vehicle**
    - **Bus**
      - Structure
      - ACS
      - Power Sys.
      - Thermal Control Sys.
      - etc.
    - **Payload**
      - Instrument System A
      - Instrument System B
      - Data Proc. & Trans.Sys.
      - etc.
  - **Ground Element**
    - **SV Control Segment**
      - Telemetry Downlink
      - Command Uplink
      - TC&C Proc. Sys.
      - etc.
    - **Data Collection Segment**
      - InstrmntData Downlink
      - Data Proc. & Rec. Sys.
      - etc.

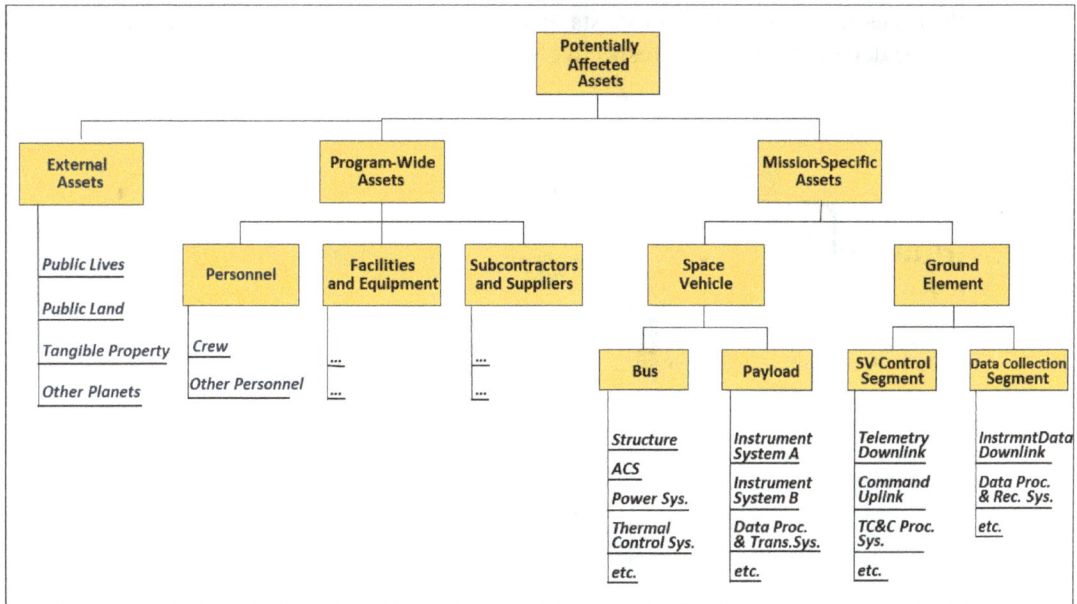

Figure 47. Example Asset Taxonomy

Taxonomies are subject to modification over time as risks are identified that suggest revisions to the categories, further partitioning of categories, or the addition of new categories. Because they integrate elements of risk that cross organizational lines, they should be maintained by the project organization. However, the project must interact with all the organizational units within it in order to ensure that the taxonomy includes all issues of importance.

When entering an individual risk into the risk database, the CONDITION, DEPARTURE, ASSET, and CONSEQUENCE should be categorized according to the appropriate risk taxonomy. The CONDITION and DEPARTURE can each be categorized according to the departure taxonomy; the ASSET can be categorized according to the asset taxonomy; and the consequence can be categorized according to the consequence taxonomy. In each case, the taxonomy is entered at the top and navigated to progressively lower levels by determining which of the nodes at the next lower level best describes the item to be categorized. This process is iterated until either the lowest level along the particular path followed is reached, or until none of the nodes at the next lower level adequately apply to the item. In the latter case, the item is categorized in the "Other" node below the last applicable node.

Over time, items may accumulate in "Other" nodes throughout the risk taxonomies. For this reason, the taxonomies should be periodically reviewed and enhanced as needed to provide appropriate categories for the items. Because the taxonomies are used to communicate risk characteristics throughout the relevant units of the NASA hierarchy (e.g., all units working under a given Program unit), modifications to the taxonomies must be coordinated among the stakeholder units and kept uniform throughout.

Figure 48 illustrates the structure of a risk statement and the application of departure and asset taxonomies to the departure and asset, respectively.

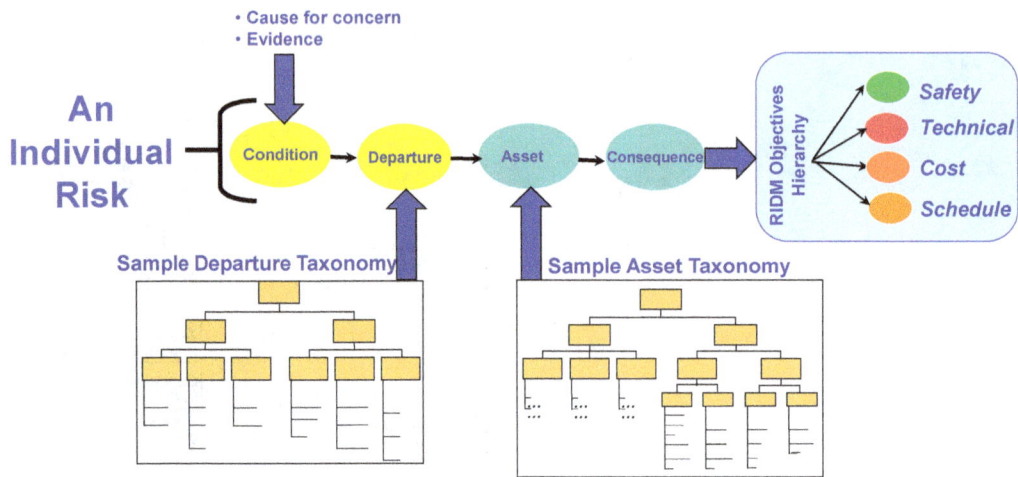

Figure 48. Risk Statement Structure and Taxonomies

The yellow box below provides an example of how the conditions and departures from a set of individual risks defined in the Identify step of CRM may be assigned to a condition/departure taxonomy. While the individual risks used in this example are summarized in the table within the yellow box, complete risk statements and narrative descriptions for them may be found in Appendix G. .

The example shows how the first attempt to fit the individual risks within the initial taxonomy may lead to the conclusion that the structure does not contain a sufficient number of categories to serve the purpose. The result is that the initial taxonomy is modified as needed (usually by a combination of expansion and rearrangement) to accommodate the individual risks without sacrificing its breadth. The final taxonomy and the assignment of the individual risks to the taxa are presented in tabular form in the example for compactness. However, a graphical form similar to Figure 46 and Figure 47 would also be appropriate.

The individual risks listed in the first table in the yellow box and described in Appendix G were selected as examples because each has some unique feature that distinguishes it from the others. For example, Risks 4(a) and 4(b) illustrate how an institutional risk, when expressed in one form, can also be a project risk when expressed in another form. Both have the same condition and departure event, but in one case the affected asset and resulting consequence are related to Center objectives and in the other case related to project objectives. The example illustrates that it is often difficult to unlink institutional and project risks, but if an institution does not manage its risks/resources well, projects can fail.

Planetary Science Mission Example: Development and Use of a Condition-Departure Taxonomy

During the initialization of CRM, the nine individual risks listed in the table below were identified and classified by condition, departure, asset, and consequence.

Number	Title	Condition	Departure	Asset	Consequence
1a	Planetary contamination	Knowledge of planet's atmosphere is limited	Higher-than-expected density causing spacecraft burn-through	Spacecraft	Planetary contamination
1b	RCS in-flight damage	Knowledge of planet's atmosphere is limited	Higher-than-expected density causing RCS damage	RCS	Inability to achieve circular orbit
2	Pu^{238} availability	The supply of Pu is low and Congress has not approved funds for processing	Cost of Pu may increase drastically	Electrical power / RTGs	Funding exceeded
3	Thrust oscillations	Launch vehicle thrust oscillations near resonance frequency of payload	Stresses may exceed design limits for instrumentation	Scientific instrumentation	Loss of scientific data
4a	DMS institutional risk	Vendor for the DB querying utility will no longer support it	Utility may become nonoperational or obsolete	Document Management System	Inability to meet data management needs of the Center
4b	DMS project risk	Vendor for the DB querying utility will no longer support it	Utility may become nonoperational or obsolete	Communications within the Project	Delay of launch date
5	Valve effect on mass margin	Valve type in RCS susceptible to corrosion	Replacement valve may be significantly heavier	RCS	Failure to meet mass requirements
6	Atmospheric sensor TRL	Atmospheric sensors are at TRL 2 and must reach TRL 8 prior to integration	The sensors may be unavailable or late	Scientific instrumentation	Failure to meet delivery date
7	Video sensor stray light	Another mission had degraded video due to stray light	Stray light may enter science package	Video system	Unacceptable loss of data quality

An attempt was made to assign these individual risks to categories in the condition-departure taxonomy of Figure 46, but it was found that many of them did not fit naturally. As a result, the condition-departure taxonomy was modified to provide a better means for classifying these individual risks. The result is shown in the table below:

Level 1 Category	Level 2 Category	Level 3 Category	Risks
External Departures	Political		
	Economic		
	Public Relations		
	Acts of God		
	Other		
Programmatic Departures	Institutional Resource Issues	Budget Resources	
		Schedule Resources	
		Info Tech & Mgmt Resources	4(a, b)
		Facility and Equip. Resources	
		Other	
	Human Resource Issues	Design Staffing	
		Software Staffing	
		Manufacturing Staffing	
		Product Assurance Staffing	
		Research & Analysis Staffing	
		Launch Support Staffing	
		Flight Operations Staffing	
		Other	
	Subcontractor & Supplier Issues	Requirement Flow-down	
		Supplied Item Defects	
		Coordination	
		Timeliness	
		Other	
	Material Resource Issues		2
	Other		

(Continued)

Level 1 Category	Level 2 Category	Level 3 Category	Risks
Technical Departures	Technology Development	Technology Readiness Level	6
		Reliability & Life Issues	
		Other	
	Design	Design Specs.	
		Performance Reqts. Unachievable	
		Margin Management	5
		Verification & Validation	
		Other	
	Fabrication	Parts Quality Control	
		Manufacturing Control	
		Manufacturing Specs.	
		Other	
	Assembly & Integration	Assembly Control	
		Assembly Specs.	
		Other	
	Testing	Testing Control	
		Test Specs.	
		Other	
	Inspection	Inspection Control	
		Inspection Specs.	
		Other	
	Operations	Operational Control	
		Operational Specs.	
		Other	
	Natural Operational Environments	Crew Environments	
		Mission/Launch Control Environs	
		Planetary/Rendezvous Environs	1(a)
		Other	
	Induced Operational Environs	Loading & Heating	1(b), 3
		Hazardous Chemicals	
		Other	7
	Other		

When more individual risks were added later on, inspection of the results revealed that they tended to congregate under two categories:

1) Programmatic departures → institutional resource issues → information technology & management resources

2) Technical departures → induced operational environments → loading & heating

This observation helped the project allocate its resources more effectively.

4.2.1.4 The Narrative Description

While the risk statement provides a concise description of the individual risk, this information is not necessarily sufficient to capture all the information that the risk identifier has to convey, nor is it necessarily sufficient to describe the concern in enough detail that risk management personnel can understand it and respond effectively to it, particularly after the passage of time. In order that enough context is recorded so that the individual risk can stand on its own and be understood by someone not otherwise familiar with the issue, a narrative description field is provided. The narrative description is format-free and can include:

- Key circumstances surrounding the "risk"

- Contributing factors

- Uncertainties

- The range of possible consequences

- Related issues such as what, where, when, how, and why.

The narrative description is also a place where the risk identifier can suggest or recommend potential mitigations or other risk responses that he/she feels is most appropriate. It is usually the case that the risk identifier is an engineer with significant subject matter expertise in the affected asset, and it is important to capture that expertise, not only concerning the nature of the issue, but also its remedy. When a risk response is recommended, the risk identifier should also record the rationale for the recommendation, preferably including an assessment of the expected risk shifting (e.g., from a safety risk to a cost risk) that would result.

An example of a narrative description for the planetary science mission is presented below. Other examples may be found in Appendix G.

Narrative Description Example for Planetary Science Mission

The atmosphere of Planet X has been observed with ground-based and Earth-orbital telescopes at various times, including during eclipses, and spectral analysis of the data has been performed. There have also been flybys to observe atmosphere thickness, species, and density. Uncertainties in the results are large because of inherent variability in the atmosphere, both spatially and timewise, making it difficult to make global inferences from limited observations. Other direct sources of uncertainty include limitations in instrument accuracy and variable solar flux effects. Additionally, there is considerable inherent uncertainty in the models used for calculating thermal responses and stresses in the heat shield, bond, and structure because these models are based on assumptions about the effects of ionizing radiation on heat transfer and the condition of the vehicle surface as it affects boundary layer transition. Associated testing in wind tunnels, hot gas facilities, and plasma arcs may be of limited applicability for the Planet X atmosphere. If the spacecraft should break up during the aerocapture maneuver, analysis shows that because of the extra-orbital velocity of the spacecraft (hyperbolic entry), it is highly likely that Pu would scatter and some fraction would reach the surface.

4.2.2 Sources of Risk Identification

Individual risks can be identified at any point in the lifecycle of a program/project, and indeed it is a fundamental principle of continuous risk management that the process of risk identification is ongoing. The primary sources of risk identification are expected to be:

- **The Initial Risk Analysis** – The initial risk analysis has its origin in the risk analysis of the selected alternative developed during the RIDM process. During RIDM, significant uncertainties with the potential to adversely affect performance were identified, and a risk list was generated identifying the major scenarios contributing to the risk of that alternative. When CRM is initialized, the RIDM risk analysis is updated to reflect risk with respect to performance requirements (as opposed to performance commitments), and additional modeling detail is added as needed in areas that were non-discriminators during RIDM and/or to adequately reflect the baselined project plan. This process may be iterative, using the risk model to assess project plan options.

  The net result is that, as discussed in Section 4.1.2.1, CRM is initialized with a risk analysis that is used going forward to identify the major scenarios contributing to performance risk. These scenarios should be captured as individual risks.

- **Taxonomy-Facilitated Brainstorming** - Brainstorming is a common method for identifying potential risks; various types of brainstorming techniques such as Checklist, What-if Analysis, Failure Modes and Effects Analysis (FMEA), and Hazard and Operability Studies (HAZOP) have been used for decades in the process industries to identify potential process upsets. Although numerous techniques are potentially applicable to the CRM *Identify* step depending on the nature of the activity, the risk taxonomies can be used as a project-specific tool for brainstorming. Specifically, each unique combination of taxa from the departure, asset, and consequence taxonomies can be used in what-if fashion to structure the brainstorming sessions and stimulate thought. For example, questions of the form, "Could a [departure taxon$_i$] acting on the [asset taxon$_j$] lead to [consequence taxon$_k$]?" can be posed for all possible combinations (or a practical set of combinations, if the total number is very large).

  An example of the use of a taxonomy to identify a new individual risk is presented in the yellow box below.

- **Conditions arising during Implementation** – As implementation proceeds, conditions can emerge that signal the presence of risk. There are a variety of potential sources of risk-indicating conditions including:

   o Data from systems engineering – As discussed in the NASA Systems Engineering Handbook [2], critical or key success or performance parameters are monitored during implementation by comparing the current actual achievement of the parameters with the values that were anticipated for the current time and projected for future dates.[19] These data are used to confirm progress and identify deficiencies that might jeopardize meeting a performance requirement, including cost and schedule constraints. When a parameter value falls outside the expected range around the anticipated value, it signals a need for evaluation and corrective action.

   o Tracking data for implemented risk responses – As discussed in Section 4.5, *The CRM Track Step*, the risk responses of *watch*, *mitigate*, and *research* entail the specification of data that will be tracked to monitor implementation of the response and assess its effectiveness in addressing performance risk. Similar to the case for data from systems engineering, when CRM tracking data fall outside expectations, it signals the need for evaluation and corrective action. If such

---

[19] Including Technical Performance Measures (TPM)

action can be accomplished within the framework of the risk response, it is not necessary to identify a new individual risk in addition to the individual risk(s) already underlying the current risk response. However, if the data are such that a new risk response is warranted, then the development of a new individual risk is advisable.

○ Inter-organizational communications – In each step of the CRM process, risk information is communicated among the various risk management organizations within the NASA hierarchy, according to the degree to which they are all working towards the accomplishment of high-level requirements and objectives, and/or are vulnerable to similar conditions and departure events (i.e., so-called "cross-cutting risks"). GIDEP alerts are a good example. This information may indicate the presence of risk within a given organizational unit that until then had been unidentified.

○ Risks elevated from lower levels in the organizational hierarchy[20] – If a unit in the NASA hierarchy is unable to adequately manage its performance risk, it may elevate the management of its risk to the unit at the next higher level of the NASA hierarchy, which is the unit that owns the requirements that are at risk. When this is the case, the situation is identified as an individual risk at the higher level if it has not already been included as such. The CRM process can then be applied in order to assess the lower level unit's performance risk in terms of its impact on the higher level unit's performance risk.

○ External sources of risk – Systems engineering is seldom accomplished in isolation from external considerations, such as the price and availability of parts and/or raw materials; the price, availability, and skill sets of human capital; or the availability of test facilities and other support functions. Project planning necessarily involves assumptions about such considerations, which, as time goes on, may prove not to be the case. The CRM *Identify* step captures such situations as individual risks when external conditions change in ways that adversely affect performance risk.

- **Rebaselining of Requirements** – It is possible that the unit at the next higher level of the NASA hierarchy will need to revise their derived requirements as part of a risk mitigation effort at its level. When this is the case, the requirements that flow down from the higher level to the current level are rebaselined in a negotiated fashion, leading to a modified set of requirements against which performance risk is assessed. As discussed in Section 2.3.2, the rebaselining may involve an adjustment process, wherein certain requirements are modified to make them more applicable and practicable, or alternatively an outright waiving of requirements that are unnecessary or counterproductive. These rebaselined performance requirements will typically require modification of the baseline project plan, which in turn will require modification of the risk model to properly reflect the rebaselined project.

---

[20] See Section 4.4.1.1 for a discussion of elevation.

The net result is that the risk analysis will require rebaselining to reflect the shift, potentially producing a spectrum of new scenarios that should be captured as individual risks.

4.2.3 Risk Advocacy and Ownership

There is no link necessary between the organizational unit or person that identifies an individual risk and the organization(s) responsible for the performance risk(s) affected. Nevertheless, it is critical to the effectiveness of the CRM process that <u>every</u> individual risk receive the appropriate level of analysis necessary to incorporate it into integrated performance risk models. So, part of the CRM Identify step is to assign each individual risk an advocate person or organization.

An advocate for an individual risk ensures that:

1. The individual risk meets the criteria for a risk and is properly recorded in the risk database.

2. The individual risk receives the appropriate level of graded analysis in a timely manner so that the effects of the individual risk are effectively reflected in relevant performance risk models.

3. The potential responses identified by the risk initiator or discovered in the analysis process (possible mitigation, or research actions) are assessed for their effect on Risk Drivers, and these potential mitigation actions and their relationships with risk drivers are recorded in the risk database to support planning and tracking.

4. The individual risk and its effect on performance risks are communicated in a timely fashion to the risk board.

5. Closure or non-closure of the individual risk occurs in accordance with the decision of the risk board.

Because the CRM process focuses on the effective mitigation of performance risks, the risk advocate's role is significantly different from that of the "traditional" risk owner. A risk advocate is <u>not</u> responsible for ensuring the individual risk is mitigated to an acceptable level, only that the risk is properly included in the integrated risk management process.

The next yellow box illustrates how each example risk introduced in the yellow box in Section 4.2.1.3 and elaborated upon in Appendix G might be assigned to an organizational unit that would act as an advocate. It also provides the rationale for those assignments.

Appendix F provides additional information about the risk ownership function in relation to other risk management functions, and about the formalization of roles and responsibilities in the RMP.

Planetary Science Mission Example: Assigning Individual Risks to the Organizational Units' Risk Lists

The risk lists of each organizational unit in the Planetary Science Mission project were developed by assigning each identified individual risk to the unit, or units, considered best able to manage it. The figure below shows their initial locations within the project's organizational hierarchy. The project itself is part of the Planetary Exploration program.

Planetary science mission project structure, requirements flowdown, and risk lists

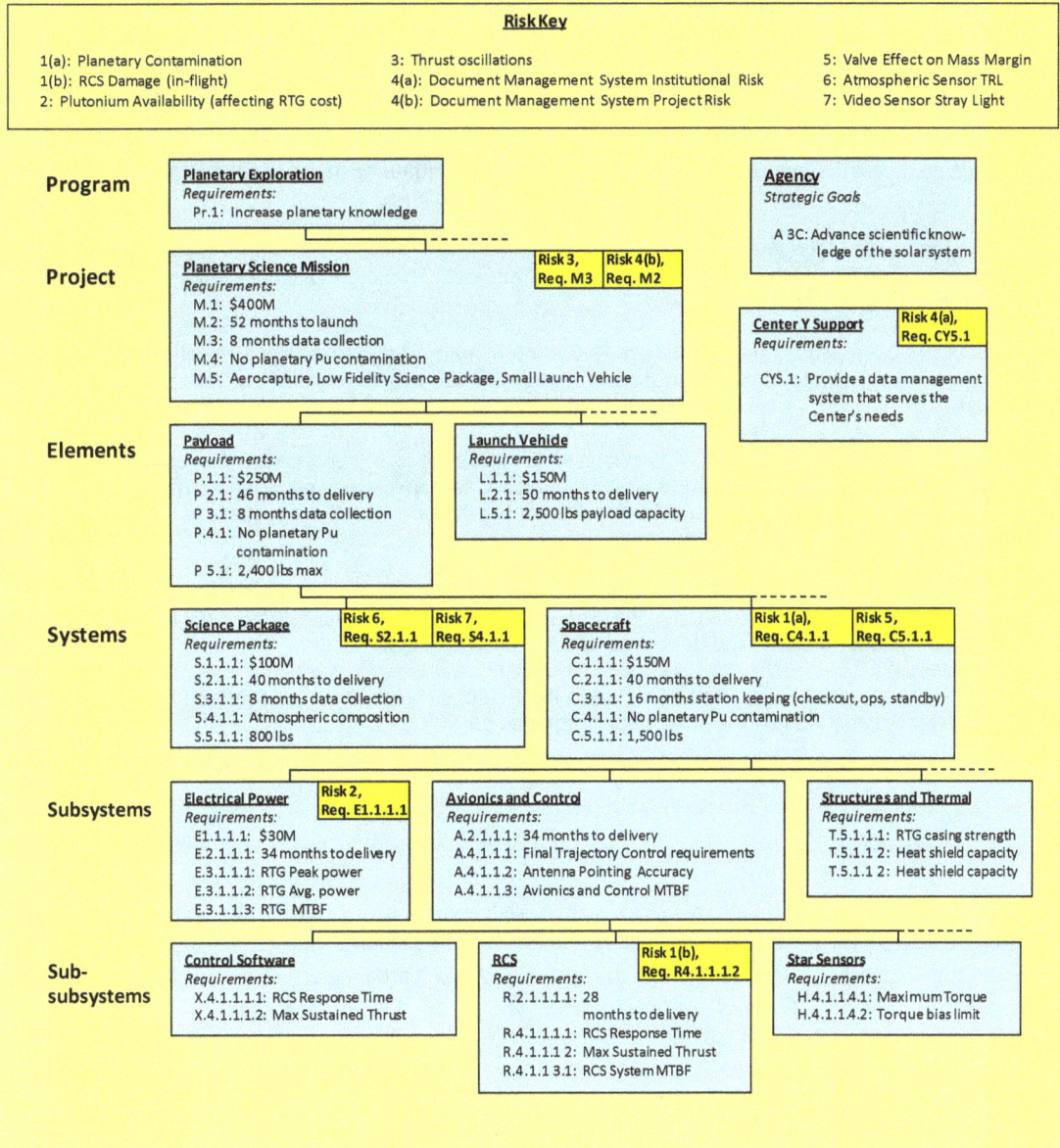

Risk Key

1(a): Planetary Contamination	3: Thrust oscillations	5: Valve Effect on Mass Margin
1(b): RCS Damage (in-flight)	4(a): Document Management System Institutional Risk	6: Atmospheric Sensor TRL
2: Plutonium Availability (affecting RTG cost)	4(b): Document Management System Project Risk	7: Video Sensor Stray Light

Program

Planetary Exploration
Requirements:
Pr.1: Increase planetary knowledge

Agency
Strategic Goals

A 3C: Advance scientific know-ledge of the solar system

Project

Planetary Science Mission
Requirements:
M.1: $400M
M.2: 52 months to launch
M.3: 8 months data collection
M.4: No planetary Pu contamination
M.5: Aerocapture, Low Fidelity Science Package, Small Launch Vehicle

Risk 3, Req. M3 | Risk 4(b), Req. M2

Center Y Support
Requirements:

Risk 4(a), Req. CY5.1

CYS.1: Provide a data management system that serves the Center's needs

Elements

Payload
Requirements:
P.1.1: $250M
P 2.1: 46 months to delivery
P 3.1: 8 months data collection
P.4.1: No planetary Pu contamination
P 5.1: 2,400 lbs max

Launch Vehicle
Requirements:
L.1.1: $150M
L.2.1: 50 months to delivery
L.5.1: 2,500 lbs payload capacity

Systems

Science Package
Requirements:
S.1.1.1: $100M
S.2.1.1: 40 months to delivery
S.3.1.1: 8 months data collection
5.4.1.1: Atmospheric composition
S.5.1.1: 800 lbs

Risk 6, Req. S2.1.1 | Risk 7, Req. S4.1.1

Spacecraft
Requirements:
C.1.1.1: $150M
C.2.1.1: 40 months to delivery
C.3.1.1: 16 months station keeping (checkout, ops, standby)
C.4.1.1: No planetary Pu contamination
C.5.1.1: 1,500 lbs

Risk 1(a), Req. C4.1.1 | Risk 5, Req. C5.1.1

Subsystems

Electrical Power
Requirements:
E1.1.1.1: $30M
E.2.1.1.1: 34 months to delivery
E.3.1.1.1: RTG Peak power
E.3.1.1.2: RTG Avg. power
E.3.1.1.3: RTG MTBF

Risk 2, Req. E1.1.1.1

Avionics and Control
Requirements:
A.2.1.1.1: 34 months to delivery
A.4.1.1.1: Final Trajectory Control requirements
A.4.1.1.2: Antenna Pointing Accuracy
A.4.1.1.3: Avionics and Control MTBF

Structures and Thermal
Requirements:
T.5.1.1.1: RTG casing strength
T.5.1.1 2: Heat shield capacity
T.5.1.1 2: Heat shield capacity

Sub-subsystems

Control Software
Requirements:
X.4.1.1.1.1: RCS Response Time
X.4.1.1.1.2: Max Sustained Thrust

RCS
Requirements:
R.2.1.1.1.1: 28 months to delivery
R.4.1.1.1.1: RCS Response Time
R.4.1.1.1 2: Max Sustained Thrust
R.4.1.1 3.1: RCS System MTBF

Risk 1(b), Req. R4.1.1.2

Star Sensors
Requirements:
H.4.1.1.4.1: Maximum Torque
H.4.1.1.4.2: Torque bias limit

Risk 1(a), *Planetary Contamination*, was assigned to the Spacecraft organization because managing it was expected to entail a coordinated effort among multiple lower-level organizations, such as Structures and Thermal, Control Software, and RCS. It was not assigned to a level higher than Spacecraft because management was expected to be done within the Spacecraft requirements (e.g., managing this risk was not assumed to entail a mass tradeoff between Spacecraft and Science Package).

Risk 1(b), *RCS Damage* in-flight, was assigned to RCS for the simple reason that its management was expected to be done within the RCS requirements.

Risk 2, *Pu^{238} Availability* affecting RTG cost, was assigned to Electrical Power because that is the unit responsible for providing long term power resources within cost.

Risk 3, *Thrust Oscillations* was added to the risk list at the Project level because it was recognized that the key issue is the lack of vibration limit requirements on the launch vehicle and the lack of vibration tolerance requirements on the payload, the development of which are the responsibility of the project. Management of this risk could not be done at a lower level, regardless of the ease of technical solution, because without well-defined requirements there is not an adequate basis for assessing the effectiveness of the solution.

Risk 4(a) *Document Management System (DMS) Institutional Risk*, was assigned to the DMS support organization in Center Y, which is separate from the project. This organizational unit is responsible for document management support systems.

Risk 4(b), *Document Management System Project Risk*, was assigned at the Project level because it affects communications throughout the project and contributes to the risk of not meeting the required launch date.

Risk 5, *Valve Effect on Mass Margin*, was assigned to the Spacecraft unit because the requirement that is most directly at risk is the spacecraft mass limit, which was considered manageable by the Spacecraft unit (i.e., it was thought to possibly entail tradeoffs among the spacecraft subsystem mass requirements, but not between, say, the spacecraft and the science package).

Risk 6, *Atmospheric Sensor TRL*, was assigned to the Science Package unit because that is the organization responsible for developing the atmospheric sensors. This risk was associated with the requirements on both schedule and data volume since each can be affected.

Risk 7, *Video Sensor Stray Light*, was assigned to the Science Package unit under the understanding that the issue is wholly produced by the science package design.

4.3   Analyze Step

The CRM *Analyze* step takes a performance risk perspective, assessing the aggregate, or cumulative, effects of individual risks on each organizational unit's performance risk. To perform this function effectively it is divided into two parts, referred to as the *Quick Look Analyze* step and the *Graded Approach Analyze* step. These two parts of *Analyze* were introduced in Section 4 and in particular in Figure 40.

The objective of the *Quick Look Analyze* step is to obtain an early ranking of the importance of each individual risk relative to the performance requirements. Importance, however, is measured from two points of view, which may at least conceptually be thought of as "tactical importance" and "strategic importance." The distinction between "tactical" and "strategic" was discussed in the introductory remarks in Section 4. Accordingly, the *Quick Look Analyze* step organizes the individual risks into two separate lists, which are called the *tactical* or *near-term criticality* list and the *strategic* or *long-term criticality* list.

The *tactical* or *near-term criticality* list ranks the individual risks in terms of their urgency with respect to initiating a near-term response. They are, in general, risk issues that can potentially cause an aggregate performance risk to change from tolerable to marginal or intolerable[21], and for which only a limited timeframe exists for instituting an effective risk management response. This list provides the basis for initiating near-term responses in the CRM Plan step.

The *strategic* or *long-term criticality* list ranks the individual risks in terms of their importance with respect to performing more detailed analysis. They are risk issues that not only can potentially cause a performance risk to change to a less tolerable state, but also have particular features that make their likelihood and/or severity inherently uncertain. This list provides the basis for prioritizing the level of detail to be pursued in performing more detailed analyses of their effects on the aggregate performance risks.

A schematic of the inputs to, tasks within, and outputs from the *Quick-Look Analyze* step is presented in Figure 49.

The objective of the *Graded Approach Analyze* step is to identify risk drivers (see the Section 4 introductory remarks for the definition of this term). These are specific elements that, because of the uncertainty attached to them, drive a performance risk to be marginal or intolerable. They can be departures that appear in one or more risk statements, events that can ensue from a departure, performance parameters that influence the likelihoods of these events, or even elements of the models themselves. The risk drivers are used to devise research and/or mitigation responses that address the strategic, or long-term, needs of risk management in the CRM Plan step.

A key facet of the *Graded Approach Analyze* step is that, similar to RIDM, the rigor of the analysis need only be enough to support robust decision making with respect to prioritization of risk issues, generation of candidate risk responses, and robust decision making from among the identified alternatives. In application of the Analyze step, it is expected that there would be ongoing communication and deliberation among stakeholders both within and among organizational units that would support a degree of parallel risk response option generation and downselection.

---

[21] The concept of marginal risk is notional for purposes of this handbook, and is intended to allow for the idea that the boundary between tolerable and intolerable may have a "fuzzy" aspect.

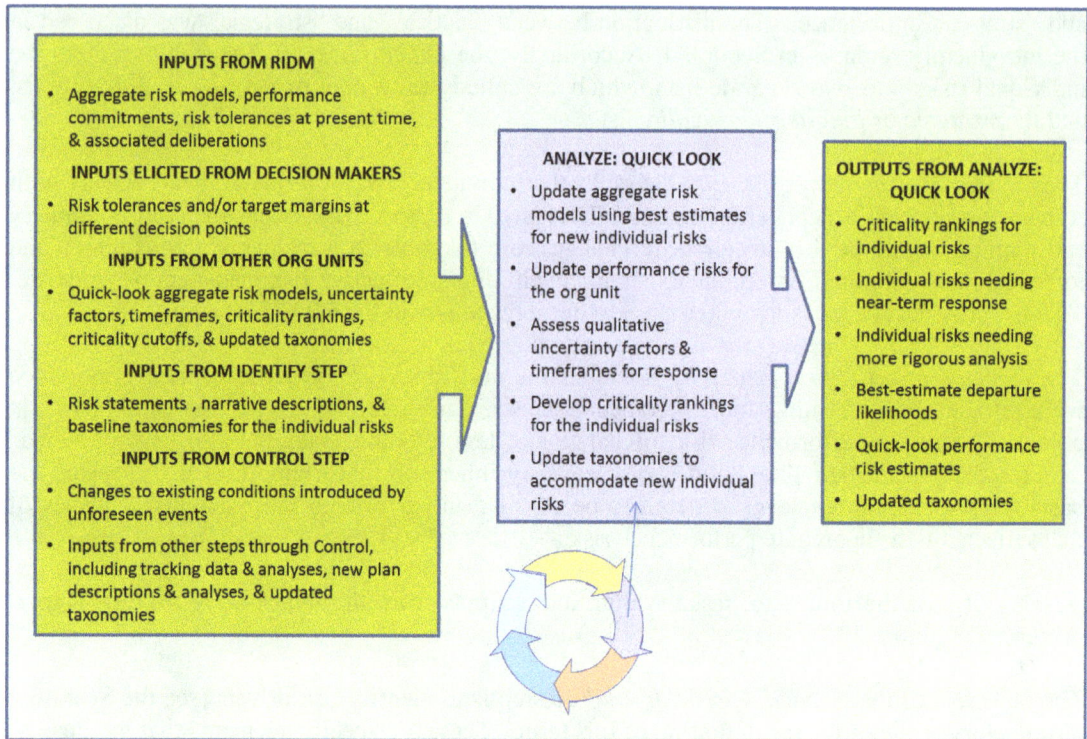

Figure 49. Quick-Look Analyze Step

A schematic of the inputs to, tasks within, and outputs from the Graded Approach Analyze Step is presented in Figure 50.

4.3.1 Introduction to Graded Analysis and the Use of Risk Scenario Diagrams in CRM

The need for a graded approach is highlighted in the following two quotations from NPR 8000.4A:

- "The resources and depth of analysis need to be commensurate with the stakes and the complexity of the [risk] being addressed."

- "The requirement to consider uncertainty is [to be] implemented in a graded fashion. If uncertainty can be shown to be small based on a simplified (e.g., bounding) analysis, and point estimates of performance measures clearly imply a decision that new information would not change, then detailed uncertainty analysis is unnecessary. Otherwise, some uncertainty analysis is needed."

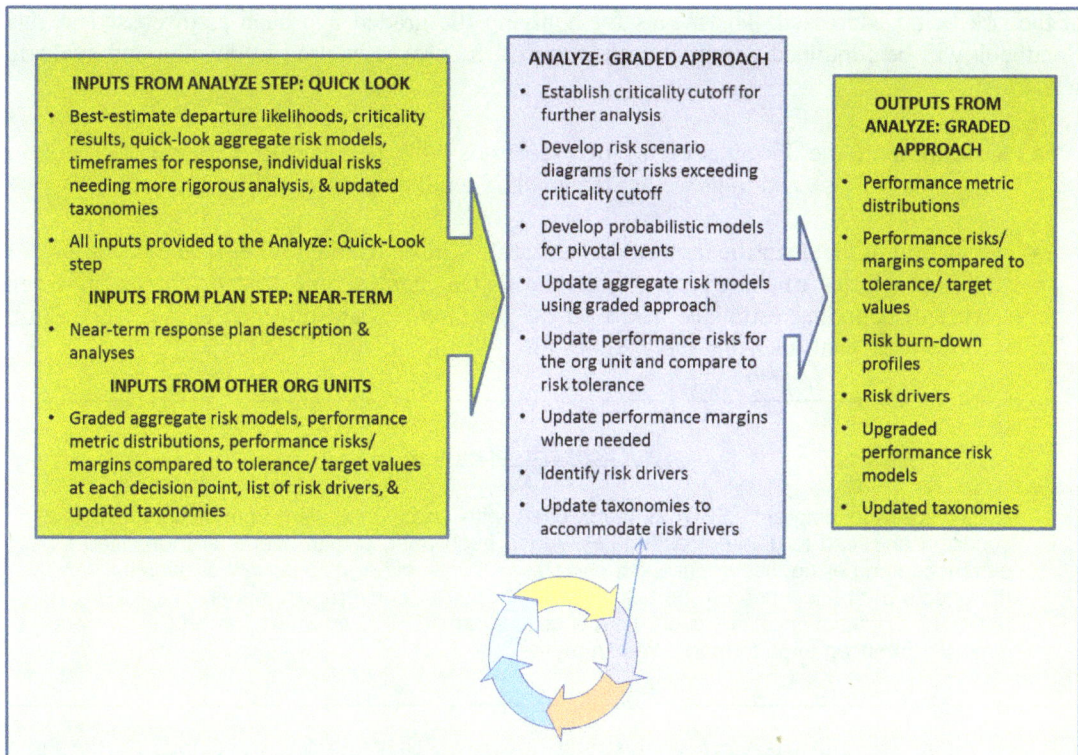

INPUTS FROM ANALYZE STEP: QUICK LOOK

- Best-estimate departure likelihoods, criticality results, quick-look aggregate risk models, timeframes for response, individual risks needing more rigorous analysis, & updated taxonomies
- All inputs provided to the Analyze: Quick-Look step

INPUTS FROM PLAN STEP: NEAR-TERM

- Near-term response plan description & analyses

INPUTS FROM OTHER ORG UNITS

- Graded aggregate risk models, performance metric distributions, performance risks/ margins compared to tolerance/ target values at each decision point, list of risk drivers, & updated taxonomies

ANALYZE: GRADED APPROACH

- Establish criticality cutoff for further analysis
- Develop risk scenario diagrams for risks exceeding criticality cutoff
- Develop probabilistic models for pivotal events
- Update aggregate risk models using graded approach
- Update performance risks for the org unit and compare to risk tolerance
- Update performance margins where needed
- Identify risk drivers
- Update taxonomies to accommodate risk drivers

OUTPUTS FROM ANALYZE: GRADED APPROACH

- Performance metric distributions
- Performance risks/ margins compared to tolerance/ target values
- Risk burn-down profiles
- Risk drivers
- Upgraded performance risk models
- Updated taxonomies

Figure 50. Graded Approach Analyze Step

Based on this guidance, Sections 3.2.1.3 and 3.2.1.4 of this Handbook have discussed the use of a graded approach for evaluating performance risks for RIDM. These sections apply for CRM as well, but there are additional considerations. The additions concern 1) the level of detail that should be exercised in developing scenarios that can lead to performance shortfalls, and 2) the rigor exercised in modeling uncertainties in the analysis of the probabilities of the events that comprise those scenarios. These considerations are especially important for CRM because:

- It is often important to know the details of the most important scenarios in order to devise the most effective mitigation strategies,

- The most significant scenarios emanating from an identified risk issue are sometimes different from the ones that are thought to be most significant at the time the risk issue is first identified,

- The significance of a scenario relative to a performance risk is often highly dependent on the uncertainty in its probability of occurrence.

These considerations suggest the need for extending the graded approach discussed earlier into the realm of scenarios. Both the formulation of the scenarios and their quantification are subject to the need for applying different levels of detail depending upon the stakes and the complexity of the risk being addressed. The means for applying the graded approach as discussed in this Handbook will be contained within two elements of the processes used to develop and evaluate scenarios:

- Gradation in the formulation of the scenarios will be handled within risk scenario diagrams (RSDs), and in particular the level of detail contained within them.

- Gradation in the quantification of the scenarios will be handled within the calculations of the probabilities of the events within the RSDs (e.g., simple best estimates or, when justified, bounding estimates based on the judgment of an informed expert, versus full probabilistic analysis using the methods discussed in Section 3.2.2.1).

Risk Scenario Diagrams

A risk scenario diagram (RSD) is a flowchart with paths that start from a unique initial condition and lead to different end states. Along each path, pivotal events are identified as either occurring or not occurring. Each end state reflects, either qualitatively or quantitatively, the effects of the scenario on the ability to satisfy performance requirements. The RSD is a particular instantiation of an event sequence diagram (ESD), in which the end states are directly referenced to performance requirements.

The example in the yellow box below illustrates how two RSDs might be developed for the same individual risk, one at a very simple level consistent with the information contained in the risk statement, and the other at a more detailed level consistent with additional information available to the risk analyst. More information on the development of RSDs will be provided in Section 4.3.3.1.

4.3.2   Quick Look Analyze Step

As mentioned in the Section 4 introductory remarks, conditions can arise that indicate individual sources of significant risk, for which only a limited timeframe exists for effective risk management response. Such situations must be responded to quickly, on a shorter time scale than that required for quantitative risk analysis, so that the window of opportunity for response is not lost.

The principal task within the quick-look analyze step is to derive two initial rankings for each individual risk: a near-term criticality ranking and a long-term criticality ranking. The near-term ranking provides the basis for deciding upon whether a near-term, quick-action tactical response plan is necessary to mitigate the individual risk, even though detailed risk analysis may not yet be completed. The long-term criticality ranking provides the bases for determining the level of detail needed later in modeling this individual risk within the aggregate performance risk models, in order to support the formulation of an overall strategic response plan.

Planetary Science Mission Example: Graded Analysis Options for Scenario Development

One of the safety risks identified for the planetary science mission example concerns the possibility that Planet X could be contaminated by plutonium. Within the requirement that there be no contamination, one of the contributing individual risks concerns the uncertainty in the density of the atmosphere surrounding the planet. The risk statement for this individual risk is as follows:

> Given that [CONDITION: the state of knowledge of Planet X's atmosphere is limited; the fact that it is difficult to ascertain more information about Planet X's atmosphere from Earth; and the fact that the spacecraft contains radioactive material], there is a possibility of [DEPARTURE: unanticipated atmospheric characteristics during the aerocapture maneuver at Planet X leading to a less-than-optimal trajectory] adversely impacting [ASSET: the spacecraft], thereby resulting in [CONSEQUENCE: spacecraft breakup and radioactive contamination of Planet X].

When the risk was first identified, it was believed that the principal issue was that the atmospheric density might be higher than expected, which could cause the spacecraft to burn up in the atmosphere during the maneuver. It was believed that if this happened, the RTG canisters would also burn up leading to a release of plutonium. A simple risk scenario diagram for this case, which contains only information consistent with the risk statement, is shown below. It consists of two pathways, one leading to a successful result (no planetary contamination) and the other leading to planetary contamination.

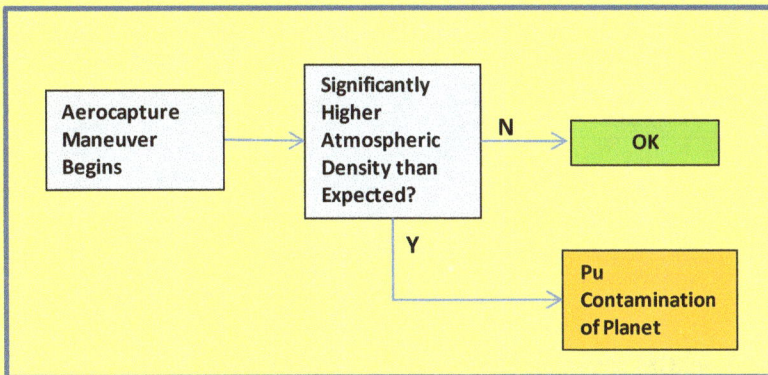

Figure 51. Simple Risk Scenario Diagram

Depending on the importance of this risk (to be addressed in a later example), It might be desirable also to consider other scenarios that could result from uncertainties in the density of the planetary atmosphere. For example, the possibility that the planet atmosphere has a lower-than-expected density might be important. If that were the case, the spacecraft might crash into the surface of the planet during the attempted aerocapture because of the lower aerodynamic drag and the need to establish a steeper trajectory to achieve the required deceleration. That could lead to a higher percentage of RTG plutonium coming into contact with the planet in a more localized area.

Another possibility is that the canisters containing the plutonium might be strong enough and sufficiently resistant to heat that they might survive the atmosphere even though the spacecraft did not. Planetary contamination would not occur in that case, but loss of mission (another performance risk) would result.

Still another possibility is that the spacecraft might survive the environmental challenges of increased heating or reduced drag, but the effectiveness of the trajectory for aerocapture might be compromised to where it might be impossible to achieve a decent orbit around the planet. This too could result in a loss of mission, or at minimum, a degradation of science because the spacecraft's distance from the planet would not be optimal for the cameras and instrumentation.

An RSD that includes these additional scenarios is shown below. Note that the number of consequential end states (denoted by the orange colored boxes) has increased from one to three. These two additional end states are accompanied by an increase in the number of pathways (or scenarios) from the initiating condition to the various end states. The number of paths has increased from two to twelve, and from Figure 52 they may be portrayed as follows:

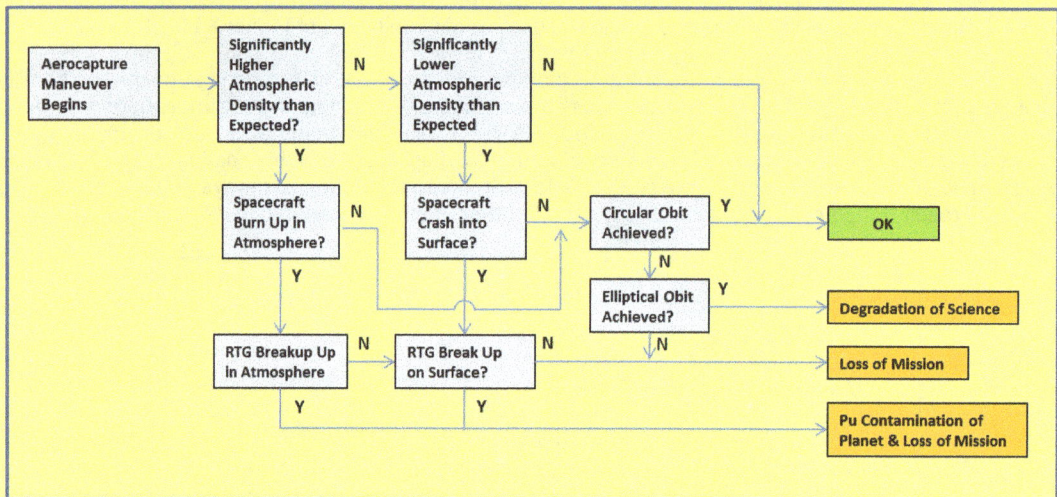

Figure 52. Moderately Detailed Risk Scenario Diagram

Pathways Leading to an "OK" end state:

1. The atmospheric density is not significantly higher than and not significantly lower than expected

2. The atmospheric density is significantly lower than expected but the spacecraft does not crash into the surface and a circular orbit is subsequently achieved

3. The atmospheric density is significantly higher than expected but the spacecraft does not burn up in the atmosphere and a circular orbit is subsequently achieved

Pathways Leading to a Degradation of Science:

4. The atmospheric density is significantly lower than expected but the spacecraft does not crash into the surface; however, the orbit subsequently achieved is elliptical rather than circular due to the deceleration being lower than optimal.

5. The atmospheric density is significantly higher than expected but the spacecraft does not burn up in the atmosphere; however the orbit subsequently achieved is elliptical rather than circular due to the deceleration being higher than optimal.

Pathways Leading to a Loss of Mission:

6. The atmospheric density is significantly lower than expected but the spacecraft does not crash into the surface; however, no orbit can subsequently be achieved due to the deceleration being very much lower than optimal. It is assumed that the spacecraft flies by the planet.

7. The atmospheric density is significantly lower than expected and the spacecraft crashes into the surface, but the RTG container does not breakup and the plutonium is contained.

8. The atmospheric density is significantly higher than expected but the spacecraft does not burn up in the atmosphere; however, no orbit can subsequently be achieved due to the deceleration being very much higher than optimal. It is assumed the spacecraft eventually falls to the planet but the RTG canister survives the impact.

9. The atmospheric density is significantly higher than expected and the spacecraft burns up in the atmosphere, but the RTG container does not breakup in the atmosphere or on the planet surface and the plutonium is contained.

Pathways Leading to Plutonium Contamination and a Loss of Mission:

10. The atmospheric density is significantly higher than expected, the spacecraft burns up in the atmosphere, and the RTG container also burns up in the atmosphere.

11. The atmospheric density is significantly lower than expected, the spacecraft crashes into the surface, and the RTG container breaks up during the crash.

12. The atmospheric density is significantly higher than expected and the spacecraft burns up in the atmosphere; however the RTG container does not burn up in the atmosphere but does eventually impact the planet surface and break up.

The more detailed level of analysis required to evaluate these additional pathways would be justified if some of the scenarios other than the one identified in the original risk statement had the potential for becoming an important determinant for whether a performance risk was tolerable, marginal, or intolerable. If that were shown to be the case, the logic contained within the RSD (or a version similar to it) would be incorporated into the modeling for that performance risk and uncertainty distributions would be developed for those events in the associated pathways that had significant uncertainty. The increased level of detail would provide a means for improving the quantitative evaluation of the performance risk and ultimately for selecting mitigation options for consideration.

The derivation of two criticality rankings for each individual risk is based on performing an evaluation of three attributes of the risk, called the likelihood and severity criticality attribute, the uncertainty criticality attribute, and the timeframe criticality attribute.[22] These are then amalgamated into the two overall criticality rankings using a Pareto approach. The three criticality attributes and the two overall criticality rankings are defined in the blue box below.

---

[22] The method for deriving criticality rankings suggested in this handbook is one of several approaches that could be used. In this approach, the likelihood and severity aspects of risk are combined into a single "likelihood and severity" criticality attribute. Another approach would be to maintain them as separate attributes. The rationale for combining them is in keeping with the quick-look nature of the criticality analysis. In the more detailed graded-approach analyze step to be discussed in Section 4.3.3, the likelihood and severity aspects of risk will be maintained as separate entities.

Criticality Attributes and Rankings

Likelihood and Severity Criticality Attribute: The likelihood that the individual risk could cause one or more performance risks to cross over one or more tolerability thresholds (i.e., from tolerable to marginal, from tolerable to intolerable, or from marginal to intolerable)

Uncertainty Criticality Attribute: The degree to which certain qualitative uncertainty factors (e.g., uniqueness, complexity, detectability), are inherent in the individual risk

Timeframe Criticality Attribute: The amount of time available before a response must be initiated

Near-term Criticality Ranking: A ranking based on amalgamating the three criticality attributes in the following order: timeframe first, likelihood second, and uncertainty third.

Long-term Criticality Ranking: A ranking based on amalgamating the three criticality attributes in the following order: likelihood first, uncertainty second, and timeframe third.

The "Quick look" step generally corresponds to using simple RSDs (like that in Figure 51) and point-estimate probabilities for the events within the RSDs to determine the criticality attributes and rankings.

4.3.2.1 Likelihood and Severity Ranking

The quick-look likelihood and severity criticality for a newly identified individual risk (or in some cases a combination of newly identified individual risks) is based on an assessment of whether or not the individual risk causes a relevant performance risk to increase beyond a tolerability threshold at any decision point along the project timeline. The approach for evaluating the likelihood and severity attribute depends in part upon whether the performance metric that is used to evaluate performance risk is discrete or continuous in nature. The difference between a discrete and continuous performance metric is defined below.

Performance Metric Types

Discrete Performance Metric: A performance metric that is defined in terms of a finite (usually small) number of possible outcomes. Loss of crew is a discrete performance metric because there are only two possible outcomes: yes or no. An associated performance requirement might be worded as follows: "The crew shall survive the mission."

Continuous Performance Metric: A performance metric that is defined on a continuous scale and therefore has an infinite number of possible outcomes. Project cost is an example of a continuous performance metric. The wording of the associated performance requirement might be as follows: "The total project cost shall not exceed XX dollars," where a particular value is substituted for XX.

4.3.2.1.1   Discrete Performance Metrics

The tasks involved in performing the quick-look assessment of the likelihood and severity criticality attribute when the performance metric is discrete are as follows:

- Access the results of the current performance risk assessments to obtain a prior value for the performance risk that is associated with the new individual risk before the effect of the individual risk is included.

- Obtain best estimates for the departure probability of the new individual risk and the probability that the performance requirement fails to be satisfied given occurrence of the departure.

- Use the estimated probabilities to update the prior value of the performance risk.[23]

- Compare the prior value of the performance risk and the updated value to the risk tolerability thresholds provided by the decision maker at each decision point.[24]

- Rank the likelihood and severity criticality attribute as "green," "yellow," or "red" according to how the individual risk causes the performance risk to cross one or more thresholds, using agreed upon rules.

Note that this is not the only means available for performing a quick-look updating of the performance risk. For example, one could alternatively insert a simplified model for the new individual risk directly into the detailed performance risk model and then calculate the performance risk with and without the new individual risk. The choice of which approach to use involves a tradeoff between the amount of fidelity needed and the time it takes to perform the analysis.

To perform the best estimate analyses mentioned above, data should be applied as follows:

- Use project-specific data as a first choice.

- Use relevant data from other projects as a second choice.

- Use indirect or surrogate data as a third choice.

- Resolve conflicting results using expert judgment elicitation.

---

[23] Note that updating the prior value of the performance risk might involve more than just adding the new departure probability to the prior value of the performance risk. If the condition specified in the new individual risk did not exist previously, it might be necessary to address how it interacts with the prior individual risks that were considered in the current performance risk assessment. If the new condition affects one or more prior individual risks, these effects should be accounted for, again using best estimate methods.

[24] Consistent with the last footnote, the prior value of the performance risk calculated in the most recent risk assessment might have to be modified to account for changes introduced by the new condition due to its interactions with prior individual risks.

Sometimes, it is desirable to estimate the probability of the departure event and the probability of failure to meet the requirement given the departure by utilizing a model that has parameters in it. For example, the probability that an instrument fails because of high temperature (the departure event) depends on the temperature in the science module. In that case, best estimates may be obtained for the parameters in the model (e.g., temperature), and the model used to derive a best-estimate departure probability.

The example in the next yellow box illustrates conceptually how these steps could be executed to determine the likelihood and severity criticality for the risk of planetary contamination caused by the atmospheric density being higher than expected.

4.3.2.1.2   Continuous Performance Metrics

The tasks involved in performing the quick-look assessment of the likelihood and severity attribute when the performance metric is continuous are slightly different from when the metric is discrete:

- Access the results of the current performance risk assessments to obtain the prior probability distribution function for the performance metric that is affected by the new individual risk.

- Obtain two or more estimates of what the performance metric would be if the departure event of the new individual risk actually occurred: for example, a best estimate (median value) and a 98% confidence estimate.

- Estimate what the updated performance metric probability distribution would look like based on the two or more point estimates.

- Compare the prior value of the performance risk and the updated value to the risk tolerability thresholds provided by the decision maker at each decision point.[25]

- Rank the likelihood and severity criticality attribute as "green," "yellow," or "red" according to how the individual risk causes the performance risk to cross one or more thresholds, using agreed upon rules.

As discussed earlier, one could alternatively insert a simplified model for the new individual risk directly into the detailed performance risk model and then calculate the performance risk with and without the new individual risk. The choice of which approach to use again involves a tradeoff between the amount of fidelity needed and the time it takes to perform the analysis.

---

[25] The footnotes under the previous discussion for discrete performance measures apply.

The example in the yellow box below illustrates conceptually how these steps could be executed to determine the likelihood and severity criticality for the risk of launch date slippage caused by the database utility software failing or becoming obsolete.

Planetary Science Mission Example for Discrete Performance Metrics: Likelihood and Severity Criticality Ranking for Planetary Contamination caused by the Atmospheric Density Being Higher than Expected

This planetary science mission example discussed in Section 4.3 is based on the risk statement in the yellow box in that section. The performance commitment adopted during RIDM, was that "the probability of radiological contamination of Planet X should be no greater than 1 in 1000." The metric corresponding to the RIDM commitment is continuous, in that probability is defined as any real number between zero and one. On the other hand, the actual performance requirement obtained from System Engineering was expressed in the following terms: "The mission shall not lead to radiological contamination of Planet X." The metric for this requirement is binary: contamination either occurs or it does not. In addition to this change, the Agency has decided for budgetary reasons to cancel a related flyby mission of Planet X that would have provided more data on the planet's atmosphere. Because of the difference between the commitment and the requirement and the cancellation of a planned flyby, the performance risk has to be reevaluated during CRM initiation.

Figure 51 applies as the basis for performing a Quick-Look analysis of the likelihood and severity criticality attribute for this individual risk. The RSD in that figure implies that there is a minimum atmospheric density needed to cause the spacecraft to burn up and release plutonium, although it does not indicate the value. Best-estimate aerodynamic heating calculations predict that the spacecraft will fail if the density is more than 80% higher than the expected value but will otherwise survive. Estimates of the density have been obtained from an earlier flyby using various instruments to measure the spectral and reflective properties of the atmosphere. Measurement uncertainties include instrument error and the statistical uncertainty associated with a limited number of observations. In addition, uncertainties in the density at any particular point at any particular time are caused by atmospheric temperature variations and other climatic conditions, including variations in solar activity. All told, it is estimated that without the flyby that has been canceled, the standard deviation, σ, in the uncertainty for the average atmospheric density that will be encountered by the spacecraft is 40% of the mean of the measured values. Assuming the uncertainty has a normal distribution, spacecraft failure corresponds to the density being at the 2σ level of the distribution[26], which has an exceedance probability of 0.023. That is, the probability of the departure event is 0.023.

Additional observations from Earth that are planned before the launch date may reduce this probability by half, to a value of about 0.011 prior to launch, but because of the difficulty of obtaining measurements and predicting what the climatic conditions will be at the time of the mission, the uncertainty cannot be reduced further under the current plan;

---

[26] The ratio of the failure density minus the mean over the standard deviation is 0,80 / 0.40 = 2.

In addition to atmospheric density uncertainties, the RIDM risk analysis had considered the risk of contamination caused by other failures in the spacecraft. These included 1) failure of the RCS subsystem during the aerocapture maneuver, resulting in an inability to exit the atmosphere, and 2) delamination of the heat shield due to earlier radiation effects on the material that bonds the heat shield to the structure. There are no significant interactions between these prior departure events and the uncertainty in the atmospheric density.

The prior analysis had indicated that the probability of planetary contamination overall was 0.008 at the time of the analysis, of which 0.003 was due to the uncertainty in the atmospheric density and 0.005 was due to the other two sources. Thus the prior probability of planetary contamination, modified so as not to include the part due to density uncertainty, was 0.005. It was further estimated that this value could be honed down to 0.001 at the time of launch as a result of existing plans to test the heat shield performance under intense radiation and improve the performance of the bond.

During the initialization of CRM, the decision maker selected tolerance thresholds for the risk of planetary contamination. The threshold for tolerable to marginal was 0.10 at the time of RIDM, subsequently decreasing systematically to a final value of 0.001 at the time of SIR. Similarly, the threshold for marginal to intolerable was 0.20 at the time of RIDM, decreasing to 0.002 at the time of SIR. In deciding upon these tolerance thresholds, the decision maker specifically stated that the burn-down in the risk of planetary contamination from 0.1 to 0.001 was attributed to the expectation that the additional planetary observations and heat shield irradiation tests currently planned between the time of RIDM and SIR would results in large reductions in the uncertainties driving the probability of planetary contamination. Based on this statement, the results from the currently planned observations and tests should not be credited in the risk calculation until they actually occur and the expected results are corroborated.

The table below shows the results of this analysis.

	Decision Point			
	RIDM	PDR	CDR	SIR
DM's tolerance threshold for contamination risk: tolerable to marginal	0.1	0.05	0.01	0.001
DM's tolerance threshold for contamination risk: marginal to intolerable	0.2	0.1	0.02	0.002
Existing contamination risk minus the risk from atmospheric density uncertainty, derived from RIDM assessment	0.005 (Tolerable)	0.003 (Tolerable)	0.002 (Tolerable)	0.001 (Tolerable)
Modified contamination risk from atmospheric density uncertainty, derived above	0.023	0.019	0.015	0.011
Updated total contamination risk	0.028 (Tolerable)	0.021 (Tolerable)	0.017 (Marginal)	0.012 (Intolerable)

The Figure below illustrates the prior and updated risks of contaminating Planet X and the tolerability of the prior and updated risks at various points along the project timeline.

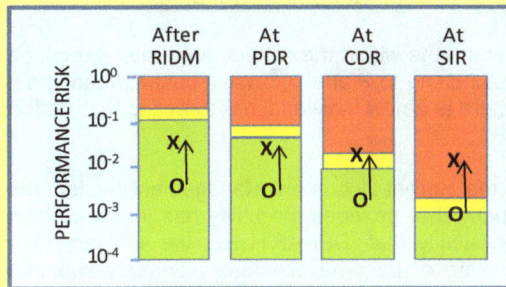

Risk of Contaminating Planet X Before and After Inclusion of Atmospheric Density Uncertainty Issue (Without Mitigation)

To determine the likelihood and severity criticality at each decision point, the following rules were used in this example:

- Designate it as low (green) if the new individual risk does not change the color of the preexisting performance risk.
- Designate it as medium (yellow) if the new individual risk changes the color of the preexisting performance risk from green to yellow.
- Designate it as high (red) if the new individual risk changes the color of the preexisting performance risk from green or yellow to red.

As the final step, a single overall likelihood and severity criticality is determined by taking the worst ranking obtained over all the decision points.

The table below shows the before-and-after tolerability levels and the likelihood and severity criticality rankings obtained from the analysis.

	Decision Point			
	RIDM	PDR	CDR	SIR
Assessed tolerability of existing contamination risk	Tolerable	Tolerable	Tolerable	Tolerable
Assessed tolerability of new total contamination risk	Tolerable	Tolerable	Marginal	In-tolerable
Likelihood and severity criticality of new risk issue	Low	Low	Medium	High
Overall likelihood and severity criticality	High			

In this case the overall criticality is determined to be red because the tolerability of the contamination risk goes from borderline tolerable (green) to intolerable (red) at SIR, The difficulty is that the risk cannot be burned down late in the project under the current plan as much as the decision maker would like.

Planetary Science Mission Example for Continuous Performance Metrics: Likelihood and Severity Criticality Ranking for Launch Date Slippage caused by Database Utility Software Failing or Becoming Obsolete

One of the performance requirements within the schedule mission execution domain is for the launch to occur before the window for launching to Planet X closes. Just five months before launch, between CDR and SIR, it was learned that there is a new individual risk affecting that performance requirement. The risk statement is as follows:

Given that [CONDITION: the current Document Management System utilizes a commercial database querying utility, the company that provides the utility has indicated they will no longer support and maintain it], there is a possibility that [DEPARTURE: the software for the utility will either reach a failed and nonrepairable state or become obsolete] adversely impacting [ASSET: communications within the Project], thereby leading to [CONSEQUENCE: a delay in the launch date beyond the launch window].

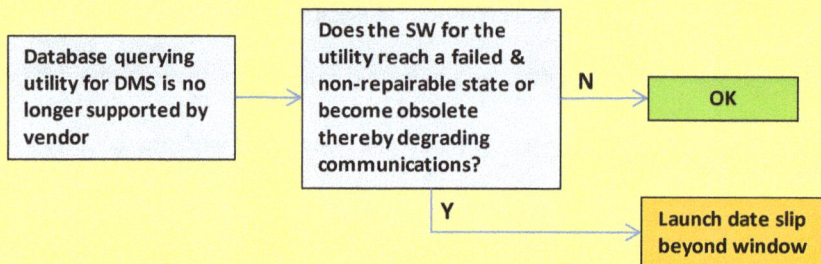

Simple Risk Scenario Diagram for Launch Date Slippage due to Obsolete Database Utility

In this example the launch date can slip by as much as 6 months without exceeding the window.

In the existing risk assessment conducted prior to this condition being identified, the risk of the launch date slipping beyond the launch window was principally caused by 1) the uncertainty associated with new technology development for one of the scientific instruments and 2) the possibility of retirements leading to a shortage of qualified engineers in the workforce. Based on past experience in the amount of time required to reach maturity for similar technology, and based on existing plans for retaining talent in essential specialties, the prior analysis of the slippage resulted in an assessment that there was 98% probability that the launch date would not slip more than 3 months beyond the scheduled date. The prior distribution for launch date slippage is shown in the illustration below as the left-most curve. It has a median value of zero and a 98% confidence value of 3 months.

The new individual risk at this stage is analyzed absent of any new planning to mitigate it. Thus, it is assumed that if the utility software fails or becomes obsolete there will be no way of repairing or replacing it with equivalent software. However, some people, it is believed, will find an ad-hoc means for interfacing a readily available database querying utility such as SQL, AQT, or even MS Word or Excel with the DMS. An assessment of similar problems that have occurred in other programs and projects, together with an elicitation of expert opinion, suggests that the launch date might be expected to slip an average (median) of one month given this occurrence, and that there is a 98% probability that it will not slip by more than three months. Its distribution is shown in the figure below as the curve with the steepest slope.

The probability distribution for the slippage due to the DMS software problem was devised already considering possible interactions between this risk and any other individual risks that could cause slippage. Therefore, to update the probability distribution for launch date slippage we need only sum the two random variables. The updated distribution has a median slippage of about 1.0 month and a 98% confidence level slippage of about 4.5 months. This distribution is also shown in the figure below as the right-most curve.

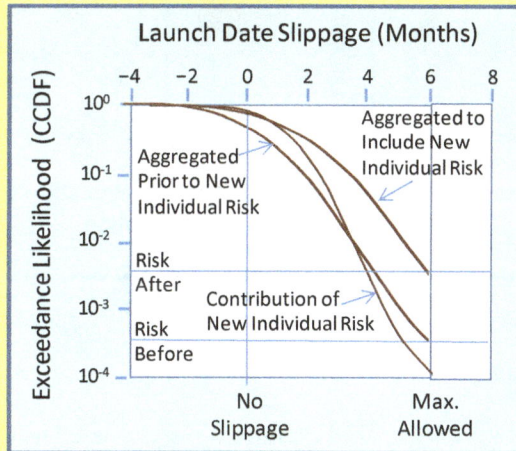

Uncertainty Distributions for Launch Date Slippage.

From these distributions, the overall probability of exceeding the 6-month window for launching was previously 0.0004 (i.e., 4×10^{-4}) but has increased to 0.004 (i.e., 4×10^{-3}) because of the new individual risk. Meanwhile, the lower tolerability threshold at SIR provided by the decision maker was 0.001 and the higher threshold was 0.005. Thus, the risk of exceeding the 6-month window was tolerable before and is marginal now. The results are summarized in the table below.

	Decision Point
	SIR
DM's tolerance threshold for missing launch window: tolerable to marginal	0.001
DM's tolerance threshold for missing launch window: marginal to intolerable	0.005
Prior launch window risk, derived from RIDM assessment	0.0004 (Tolerable)
Updated launch window risk, derived above	0.004 (Marginal)
Likelihood and severity criticality of new individual risk	Medium

The rules used in this example to determine likelihood and severity criticality are assumed to be the same as in the previous example. The figure below illustrates the prior and updated risk of not meeting the launch window requirement, and the tolerability of the prior and updated risks.

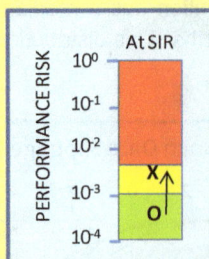

Risk of Not Meeting Launch Window Before and After Identification of the Database Utility Software Issue (Without Mitigation)

4.3.2.2 Uncertainty Ranking

Because the likelihood and severity criticality attribute is evaluated using best estimates, there is a need to account for uncertainties as part of a separate criticality attribute. At this stage of the analysis, uncertainties are considered in terms of a set of qualitative, or subjective, factors. These factors are intended to capture general characteristics of the risk that would be expected to contribute to uncertainty, although the specific amount of uncertainty associated with each factor for the individual risk in question has not yet been established.

The factors to be considered are established by the project in the RMP. An example of uncertainty factors that would be appropriate for many projects is provided in Table 4 below:

Table 4. Examples of Generic Factors for Uncertainty Ranking

Uncertainty Factor	Associated Question to be Answered
Uniqueness	Is this risk issue unique or new compared to risks that have occurred in other projects?
Cross-Cutting Character	Does this risk issue affect a large number of functions, hardware elements, software elements, or procedures and/or have the potential to cross organizational lines?
Complexity	Does this risk issue involve complex interactions between or among hardware elements, software elements, organizations, and/or individuals?
Propagation Potential	Could this risk issue lead to a propagation of events that could result in more severe consequences than the immediate events caused by the risk?
Detectability	Is there anything that inhibits the ability to detect the full extent of the risk and track its progress?

The thesis of the uncertainty ranking is that an answer of yes to any of the above questions implies uncertainty in the value of the performance metric that is affected by the individual risk in question.

Any organizational unit affected by the individual risk must also decide how to rank the uncertainty criticality attribute as red, yellow, or green. The criteria for ranking, for example, might be as follows:

- Rank the uncertainty attribute as "red" if two or more of the uncertainty attributes are answered "Yes"

- Rank as "yellow" if one of the uncertainty attributes is answered "Yes" and the others are answered "No"

- Rank as "green" if all of the uncertainty attributes are answered "No"

Planetary Science Mission Example: Uncertainty Criticality Ranking for Planetary Contamination Caused by the Atmospheric Density Being Higher than Expected

The table below lists a Yes or No answer for each uncertainty factor that might affect the risk of planetary contamination caused by the atmospheric density being higher than expected, and provides the rationale for each choice:

Uncertainty Factor	Applicable to the Individual Risk?	Rationale
Uniqueness	Yes	Although aerobraking has been used in previous missions as a means for slowing down a spacecraft, this is the first mission to attempt aerocapture (wherein the spacecraft achieves orbit around the planet). Aerocapture requires greater deceleration in the planetary atmosphere and is more demanding on the spacecraft.
Cross-Cutting Character	Yes	The loads on the spacecraft during aerocapture could affect any subsystem that has components that are external to the spacecraft, including the thermal protection system (heatshield), the reaction control system (thrusters), some of the instrumentation. Thus, the effects of atmospheric density uncertainty cross several organizational lines.
Complexity	Yes	The success of the aerocapture maneuver depends on complex physics that are affected not only by uncertainties in atmospheric density but also uncertainties in the chemical composition of the atmosphere. For example, there is a potential for chemical reactions between the atmospheric species and the external materials of the spacecraft that can affect aerodynamic heating and the strength of materials. In addition, any design to mitigate the effects of this risk would necessarily involve complex tradeoffs involving several organizational units.
Propagation Potential	No	The consequences already identified are as severe as it gets.
Detectability	Yes	The full extent of atmospheric density uncertainties is difficult to assess because it is not possible beforehand to get a comprehensive view of the entire atmosphere. Large portions of the atmosphere will not have been viewed prior to the mission.

Based on the criteria in the text, the uncertainty ranking is red because four of the five factors have been identified as being applicable to this risk.

Planetary Science Mission Example: Uncertainty Criticality Ranking for Launch Date Slippage Caused by Database Utility Software Failing or Becoming Obsolete

The table below lists a Yes or No answer for each uncertainty factor that might affect the risk of launch date slippage caused by database utility software becoming failed or obsolete, and provides the rationale for each choice:

Uncertainty Factor	Applicable to the Individual Risk?	Rationale
Uniqueness	No	Failures of support software and the threat of obsolescence have affected many projects, so there is very little that is unique about this risk.
Cross-Cutting Character	Yes	Failure or inefficiencies in the data management system would affect virtually every organizational unit that uses the system or that interacts with units that use the system.
Complexity	No	The risk as stated in the risk statement is straightforward. Mitigation does not require unusual actions.
Propagation Potential	Yes	There is a potential for the degraded state of communication to lead to unforeseen human errors. Since the project requires a large amount of cross-organizational interfacing, the number of human errors that can result from lack of effective communication is unfortunately almost limitless.
Detectability	No	If the software were to not function as intended, the effects of this nonfunctional state would be known almost immediately because people would not be able to access the data they need.

Based on the criteria in the text, the uncertainty ranking is red because two of the five factors have been identified as being applicable to this risk.

4.3.2.3 Timeframe Ranking

The timeframe of an individual risk is denoted as near-term, mid-term, or far-term based on the minimum amount of time that has to elapse before the departure event could plausibly occur. This time is compared to an estimate of the minimum and maximum amounts of time it could take to implement a mitigation response. The criteria for ranking the timeframe criticality are as follows:

- An individual risk is near-term if the minimum time for the departure event is less than or about the same as the estimated minimum time it would take to devise and implement a mitigation plan.

- An individual risk is mid-term if the minimum time for the departure event is greater than the estimated minimum time and less than the estimated maximum time it would take to implement a mitigation plan.

- An individual risk is far-term if the minimum time for the departure event is greater than or about the same as the estimated maximum time it would take to devise and implement a mitigation plan.

The timeframe ranking is red if it is near-term, yellow if it is mid-term, and green if it is far-term.

Planetary Science Mission Example: Timeframe Criticality Ranking for Planetary Contamination Caused by the Atmospheric Density Being Higher than Expected

Since the risk of planetary contamination caused by the atmospheric density being higher than expected is identified at the beginning of the project, and since the likelihood and severity criticality has been evaluated to be tolerable until near launch time, the timeframe criticality of the individual risk is far-term.

Planetary Science Mission Example: Timeframe Criticality Ranking for Launch Date Slippage Caused by Database Utility Software Failing or Becoming Obsolete

Since the risk of launch date slippage caused by the database utility software failing or becoming obsolete is identified at only five months before scheduled launch, and since the likelihood and severity criticality has been evaluated to be marginal at the next decision point, the timeframe criticality of the individual risk is near-term.

4.3.2.4 Near-Term (Tactical) Criticality Ranking

The near-term, or tactical criticality ranking is intended to serve as a means for determining which if any of the individual risks need a rapid response. Such a response is called tactical because it focuses on mitigating the individual risk before the opportunity to do so is lost.

The tactical ranking is compiled by sorting the possible combinations of criticality attribute rankings in the following order: timeframe first, likelihood and severity second, and uncertainty third. The 27 resulting ranks (3 x 3 x 3) may be condensed somewhat because those at the bottom are not as important to discriminate between as those at the top; therefore, some of the categories near the bottom can be weeded out without loss of usefulness. Table 5 below shows one way of condensing the list of 27 to 18.

Table 5. Condensing the Tactical Criticality Ranking

Tactical Rank	Timeframe	Likelihood & Severity	Uncertainty
1	Red	Red	Red
2	Red	Red	Yellow
3	Red	Red	Green
4	Red	Yellow	Red
5	Red	Yellow	Yellow
6	Red	Yellow	Green
7	Red	Green	Red
8	Red	Green	Yellow
9	Red	Green	Green
10	Yellow	Red	Red
11	Yellow	Red	Yellow
12	Yellow	Red	Green
13	Yellow	Yellow	Red
14	Yellow	Yellow	Yellow
15	Yellow	Yellow	Green
16	Yellow	Green	Red
17	Yellow	Green	Yellow
18	Yellow	Green	Green
19	Green	Red	Red
20	Green	Red	Yellow
21	Green	Red	Green
22	Green	Yellow	Red
23	Green	Yellow	Yellow
24	Green	Yellow	Green
25	Green	Green	Red
26	Green	Green	Yellow
27	Green	Green	Green

Tactical Rank	Timeframe	Likelihood & Severity	Uncertainty
1	Red	Red	Red
2	Red	Red	Yellow
3	Red	Red	Green
4	Red	Yellow	Red
5	Red	Yellow	Yellow
6	Red	Yellow	Green
7	Red	Green	Red
8	Red	Green	Yellow
9	Red	Green	Green
10	Yellow	Red	Red
11	Yellow	Red	Not Red
12	Yellow	Yellow	Red
13	Yellow	Yellow	Not Red
14	Yellow	Green	Red
15	Yellow	Green	Not Red
16	Green	Red	Any
17	Green	Yellow	Any
18	Green	Green	Any

Planetary Science Mission Example: Near-Term (Tactical) Criticality Ranking for the Two Example Individual Risks

As shown below, based on the rankings of the criticality attributes in the previous yellow boxes, the risk involving launch date slippage due to the database utility software becoming failed or obsolete ranks in Category 3 among the tactical ranks. The risk involving planetary contamination due to atmospheric density being higher than expected ranks in Category 16. The former demands a near-term tactical response much more than the latter.

Tactical Rank	Timeframe	Likelihood & Severity	Uncertainty	Individual Risks
1	Red	Red	Red	
2	Red	Red	Yellow	
3	Red	Red	Green	
4	Red	Yellow	Red	Launch date slipped due to database utility software becoming failed or obsolete
5	Red	Yellow	Yellow	
6	Red	Yellow	Green	
7	Red	Green	Red	
8	Red	Green	Yellow	
9	Red	Green	Green	
10	Yellow	Red	Red	
11	Yellow	Red	Not Red	
12	Yellow	Yellow	Red	
13	Yellow	Yellow	Not Red	
14	Yellow	Green	Red	
15	Yellow	Green	Not Red	
16	Green	Red	Any	Planetary contamination due to atmospheric density being higher than expected
17	Green	Yellow	Any	
18	Green	Green	Any	

4.3.2.4.1 Use of Near-Term Criticality Ranking to Determine Response Priorities

Any tactical ranking that involves a red or yellow timeframe is a candidate for a rapid response, but the order of urgency is dictated by the specific rank within that group. Thus, those in tactical rank 1 would receive immediate attention normally leading to as rapid a response as possible. Those in tactical rank 2 would also receive immediate attention, but the rapidity of the response would depend upon the resources available after those in tactical rank 1 had been addressed. If all the individual risks in tactical rank 2 could be responded to immediately within the resources available to the organizational unit, then the question of whether or not to initiate an immediate response defers to tactical rank 3, and so on down the list.

Below a certain cut-off tactical ranking, no rapid response would be necessary. The decision maker of each organizational unit determines where that cut-off is. For reasons of practicality, the total number of individual risks to be handled on a near-term basis would affect the decision for where the cut-off should be.

For the risks in tactical ranks above the cut-off that cannot be responded to immediately because of resource limitations, several options are possible:

- Additional resources may be provided to the responsible organizational unit from elsewhere within the project, the program, or the Agency.

- The individual risk may be elevated to a higher organizational unit where more resources are available.

- The response to the individual risk may be put on hold until resources become available within the responsible organizational unit.

4.3.2.5 Long-Term (Strategic) Criticality Ranking

The purpose of the long-term, or strategic criticality ranking is to determine the level of rigor needed in the modeling of each individual risk within the aggregate performance risk model. This ranking is called strategic because it focuses on the development of integrated models whose purpose is to help determine an integrated long-term response plan.

The strategic ranking is compiled by sorting the possible combinations of criticality attribute rankings in the following order: likelihood and severity first, uncertainty second, and timeframe third. As with the tactical ranking scheme, the 27 resulting ranks may be condensed somewhat because those at the bottom are not as important to discriminate between as those at the top.

Planetary Science Mission Example: Long-Term (Strategic) Criticality Ranking for the Two Example Individual Risks

As shown below, based on the rankings of the criticality attributes in the previous yellow boxes, the risk involving planetary contamination due to atmospheric density being higher than expected ranks in Category 3 among the strategic ranks. The risk involving launch date slippage due to the database utility software failing or becoming obsolete ranks in Category 10. The effect of the former on the aggregate performance risks demands more robust modeling much more so than the latter.

Strategic Rank	Likelihood & Severity	Uncertainty	Timeframe	Individual Risks
1	Red	Red	Red	
2	Red	Red	Yellow	
3	Red	Red	Green	Planetary contamination due to atmospheric density being higher than expected
4	Red	Yellow	Red	
5	Red	Yellow	Yellow	
6	Red	Yellow	Green	
7	Red	Green	Red	
8	Red	Green	Yellow	
9	Red	Green	Green	
10	Yellow	Red	Red	Launch date slipped due to database utility software becoming failed or obsolete
11	Yellow	Red	Not Red	
12	Yellow	Yellow	Red	
13	Yellow	Yellow	Not Red	
14	Yellow	Green	Red	
15	Yellow	Green	Not Red	
16	Green	Red	Any	
17	Green	Yellow	Any	
18	Green	Green	Any	

4.3.2.5.1 Use of Long-Term Criticality Ranking to Make Graded Analysis Decisions

Any strategic ranking that involves a red or yellow likelihood is a candidate for a more detailed modeling of the effect of the individual risk on the aggregate performance risk. Similar to the case for tactical risks, however, the level of modeling is dictated by the specific rank within that group. Thus, those in strategic rank 1 would normally require detailed modeling to include comprehensively derived RSDs and probabilistic treatment of the uncertainties. Those in strategic rank 2 would also require a more detailed level of modeling than has been done in the quick-look analysis, but the degree of rigor would be proportioned according to the amount of resources available to perform detailed modeling. The same considerations apply recursively down the list of strategic ranks.

In practice, the decision maker for each organizational unit would select a cutoff ranking below which the quick-look best-estimate analysis is considered to be sufficient and more detailed modeling is not required. All individual risks below the cut-off would be incorporated into the aggregate performance risk models to ensure completeness, but level of analysis for these individual risks would be limited to simplified RSDs and best-estimate probabilities. The level of the cutoff ranking depends upon how many risks the organizational unit is responsible for and how many can be reasonably modeled in more detail within the resources that are available.

4.3.2.6 Relationship Between Criticality Rankings and the Risk Matrix

The use of some form of "risk matrix," within which individual risks are binned, has been and still is a popular and common risk display practice. A selective and prudent use of a matrix display can be effective as a simple, supplementary communication tool for presenting a summary overview of the relative importance of individual risks.

As is mentioned in Section 6.4.2.3 of the Systems Engineering Handbook [2], risk matrices in their currently used 5x5 format have the following four limitations:

1. They do not have the ability to deal with aggregate risks (i.e., total performance risks).

2. Interaction between individual risks is not considered. Each risk is mapped onto the matrix individually. (These individual risks can be related using failure modes and effects criticality analysis (FMECA) or a fault tree.)

3. The risk matrices do not have the ability to represent uncertainties. A risk is assumed to exist within one likelihood range and consequence range, both of which are assumed to be known.

4. Tradeoff between likelihood and consequence is fixed. Using the standardized 5x5 matrix, the significance of different levels of likelihood and consequence are fixed and unresponsive to the context of the program.

To avoid the possibility of conveying misleading results when populating a risk matrix, it is essential for the entries in the matrix (i.e., the assignment of individual risks to cells within the matrix) to be based on a rational, objective, and defensible process. One way for doing this is to ensure that the entries on the matrix are consistent with the near term and long term criticality rankings developed earlier in Section 4.3. Since the criticality rankings are based on rational, objective, and defensible analysis, the matrix itself can be claimed to have these properties if it is based on the criticality rankings.

The process for deriving criticality rankings for individual risks corrects the four limitations cited above in the following ways:

1. The degree to which the affected aggregate risk changes from tolerable to intolerable based on the addition of a new individual risk is contained within the likelihood and severity criticality attribute developed in Section 4.3.2.1

2. Interactions between individual risks are also accounted for within the likelihood and severity criticality attribute, since the aggregate risk models account for identified interactions

3. Uncertainties are accounted for within the uncertainty criticality attribute developed in Section 4.3.2.2, which queries a set of qualitative factors that are indicators of uncertainty

4. The criticality rankings are responsive to the context of the program since they depend on the risk tolerances provided by the decision makers

For the matrix to be consistent with the criticality rankings, therefore, it is merely necessary for the individual risks to appear within the red, yellow, and green regions of the matrix in the same ordering as obtained from the criticality rankings. This suggests that there should be a correspondence between cells in the matrix and criticality rankings. One example for doing this is shown in Figure 53. (The specific criticality rankings that are assumed to apply to each cell may be different from those shown in the figure, as suits the analyst's judgment, so long as the relative ordering is preserved.)

Consequence Scale →

Probability Scale ↑

Consequence Index / Probability Index	1	2	3	4	5
5	Crit, 13	Crit, 12, 11, 10, 9, 8, 7	Crit, 6, 5, 4	Crit, 3, 2	Crit, 1
4	Crit, 14, 13	Crit, 12, 11, 10	Crit, 9, 8, 7	Crit, 6, 5, 4	Crit, 3, 2
3	Crit, 15, 14, 13	Crit, 12	Crit, 12, 11, 10	Crit, 9, 8, 7	Crit, 6, 5, 4
2	Crit, 17, 16	Crit, 15, 14, 13	Crit, 13	Crit, 12, 11, 10,	Crit, 9, 8, 7
1	Crit, 18	Crit, 17, 16	Crit, 15, 14	Crit, 13	Crit, 12, 11, 10

Figure 53. Example Use of Criticality Analysis to Justify the Placement of Individual Risks on a Risk Matrix

The following principles apply in constructing risk matrices from criticality rankings:

- There should be a separate risk matrix for each performance requirement that is of interest to the decision maker.
- For some performance measures, there should be a risk matrix for near-term criticality and another one for long term criticality.
- The population of the cells with individual risks should be consistent with the criticality ranking of each individual risk for the performance measure in question.
- In cases where it is possible to place an individual risk in more than one cell because the same criticality ranking applies to multiple cells, the selection between the cells should be based on whether the new risk has a larger effect on the likelihood of the departure event or the severity of the consequence.

- Once the separate risk matrices have been developed and populated with different individual risks, it is permissible to combine them into a single risk matrix for purposes of summarization, so long as the process for combining them is consistent and rational.

4.3.3   Graded Approach Analyze Step

As mentioned in the Section 4 introductory remarks, the graded approach analyze step is intended to take a more long-term, strategic view of the aggregated performance risks, in contrast to the quick-look analyze step which focuses upon obtaining a near-term, tactical view of the individual risks.

There are three principal tasks to be performed within the graded approach analyze step:

- Update the modeling and analysis for each performance risk to include all new individual risks that are open, including those that are deemed to not require detailed analysis.

- Increase the level of robustness in the modeling of individual risks that have a high enough strategic criticality ranking to justify it without exceeding available resources.

- Determine the performance parameters and/or pivotal events that most significantly drive each performance risk because of their importance to the model and the magnitude of their uncertainties.

Implementation of the second bullet generally corresponds to using more complex RSDs (like that in Figure 52) and full uncertainty distributions for input variables. Implementation of the third bullet requires sensitivity analysis to determine which input variables are driving the aggregated performance risks.

4.3.3.1 Developing Risk Scenario Diagrams (RSDs)

RSDs are the basis for more rigorously analyzing risks that are above the cut-off strategic criticality rank and are particularly useful for identifying important scenarios that otherwise might be missed. The RSDs themselves are primarily intended to be communications aids to facilitate inter-organizational discussions of the risks. They do not necessarily translate exactly to the more detailed risk analysis modeling, but they do serve as a tool for helping ensure that the models cover all the important elements of the risk.

The idea of using various levels of detail for RSDs to facilitate a graded analysis approach was introduced in Section 4.3.1. The more detailed RSDs differ from the simplest ones in several ways:

- They typically contain multiple pivotal events.

- Each pivotal event typically contains multiple levels.

- Each pathway through the various levels of the pivotal events is considered to be a scenario that leads to an end state.

- The end state may consist of one or more consequences affecting one or more assets of interest to the organizational unit.

- Each consequence and associated asset is relatable to a specific performance requirement that the organizational unit is supposed to meet.

At this stage of CRM, the intent of the RSDs is to identify the possible consequences and the pathways that can lead to them, but not yet to quantify them. Furthermore, they do not yet contain any reference to mitigation or research options that are not part of the present plan.

A generic format for RSDs is shown in Figure 54. Note that each pathway is specifically identified with one or more performance requirements (PRs) that pertain to the responsible organizational unit. For example, Pathway 7 in the figure is associated with PR 1, 2, and 3. Each performance requirement is specifically defined, and each is relatable to a specific quantifiable performance metric. The performance metric is the basis for defining the asset and the consequence. For example, PR 1 may refer to the amount or the probability of planetary contamination, PR 2 to the launch date, and PR 3 to the total cost of the project. Pathway 1 in the figure is benign (it affects none of the performance metrics), whereas Pathway 2 affects only PR 1 and Pathway 3 affects all three.

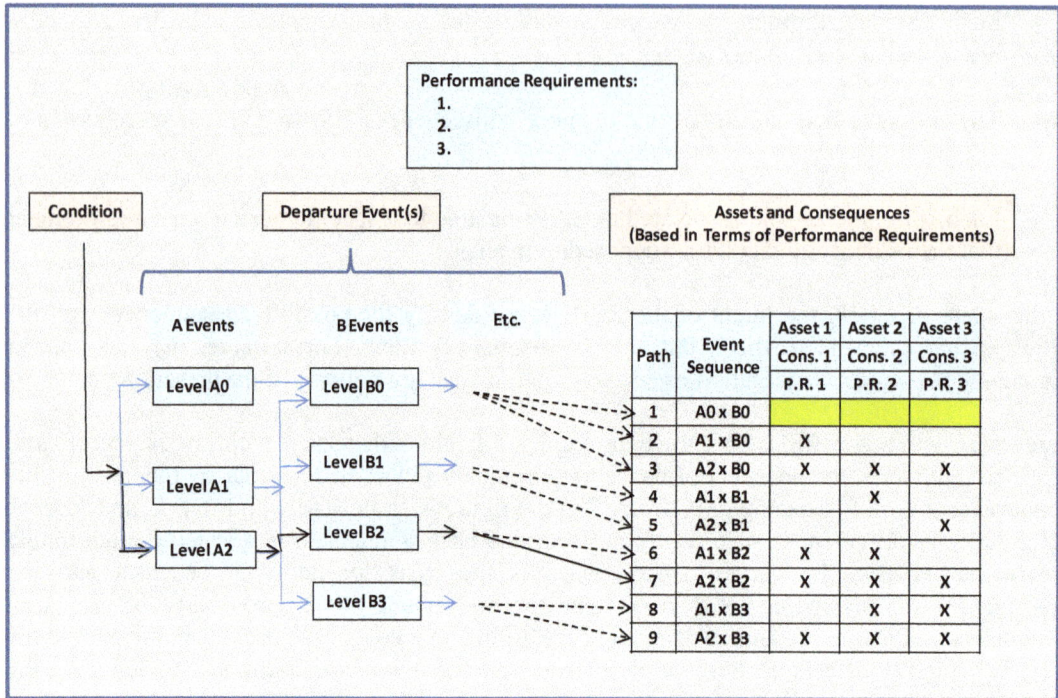

Figure 54. Format for an RSD Showing the Effects of Each Pathway on the Organizational Unit's Performance Requirements

The RSD is intentionally kept relatively simple so as to facilitate the development of scenarios and provide a clear visualization of the sequence of events leading to the consequences. Thus, not all the pathways need be shown uniquely in the RSD if they can be inferred. It is perhaps easiest to illustrate this in Figure 54 by using an analogy with a water distribution system. Suppose the box labeled "Condition" were a reservoir with a water pump and the various lines with arrows were open ducts. Because of the interconnectivity of the ducts, water could pass from the reservoir to level B0 via Level A0, Level A1, or Level A2. Similarly, water could pass from the reservoir to level B1 via Level A1 or Level A2, and so on. All told, there are nine possible water pathways. Though the pathways are not all individually drawn, each can be inferred. For example, Pathway 7 is distinctly identified by the solid black line starting from the reservoir and proceeding to Level B2 by way of Level A2. The duct analogy serves the purpose of reducing the complexity of the RSD while making it easy to account for all the pathways.[27]

The next yellow box provides an example RSD derived for Risk 1(b) in Appendix G: RCS damage caused by unexpected atmospheric density.

---

[27] Note that the RSD can be readily expanded into an event tree or event sequence diagram that explicitly shows all the pathways

Planetary Science Mission Example: Risk Scenario Diagram for the Reaction Control System being Damaged because the Atmospheric Density is Higher than Planned

This example is similar to the one considered in Section 4.3.1, but the consequence relates specifically to the reaction control system (RCS) rather than to the spacecraft as a whole. The responsible organizational unit is the RCS unit rather than the Spacecraft unit. The risk statement is as follows:

Given that [CONDITION: the state of knowledge of Planet X's atmosphere is limited; the fact that it is difficult to ascertain more information about Planet X's atmosphere from Earth; and the reaction control systems fielded to date have not needed to operate in such harsh (hyperbolic entry) environments], there is a possibility of [DEPARTURE: unanticipated atmospheric characteristics during the aerocapture maneuver at Planet X] adversely impacting [ASSET: the exposed RCS components], thereby leading to [CONSEQUENCE: damage to the RCS system making it unable to perform orbital maneuvers].

The consequence of principal interest to the RCS organizational unit is damage to the RCS that could cause it to become unable to achieve a circular orbit following the aerocapture maneuver. By contrast, the consequence of principal interest to the spacecraft organization was planetary contamination.

During the criticality ranking process, it has been determined that this individual risk has a likelihood ranking of Red, an uncertainty ranking of Red, and a timeframe ranking of Green, the same as for the risk to the Spacecraft considered in Section 4.3.2.1.1. Its overall near-term (tactical) criticality ranking is 16 and its long-term (strategic) criticality ranking is 3, the same as for the Spacecraft risk. Because of its high ranking in the strategic area, it is decided that a more detailed RSD is required in order to capture all the scenarios that might turn out to be important.

The organizational unit is assumed to have three performance objectives (although in reality it would probably have many more.) These concern the ability to achieve a circular orbit after the aerocapture maneuver (a technical requirement), the project schedule, and the project cost. Since the individual risk does not reference project schedule or cost, and since mitigation is not yet incorporated into the RSD, the requirements pertaining to schedule and cost are not germane at this point.

Because of the high strategic criticality ranking, it is decided that careful attention has to be paid to how the uncertainties pertaining to atmospheric density affect the damage amount, and how uncertainties in the damage amount affect the ability to achieve a circular orbit. Correspondingly, it is decided to consider five levels of atmospheric density deviation from the nominal and four levels of RCS damage. The five levels of density deviation are none, lower than planned, higher than planned a relatively minor amount (≤ 25%), higher than planned by a moderate amount (25% to 50%), and higher than planned by a major amount (≥50%). Correspondingly, the four levels of damage are none, minor, moderate, and major. The extent of damage implied by each of these damage levels has been defined previously by the RCS group. The resulting RSD for this example is shown in Figure 55.

Risk 1(b): RCS Damage during Aerocapture (Owned by RCS Org Unit)

Given that [CONDITION: the state of knowledge of Planet X's atmosphere is limited; the fact that it is difficult to ascertain more information about Planet X's atmosphere from Earth; and the reaction control systems fielded to date has not needed to operate in such harsh (hyperbolic entry) environments] ...

Performance Requirements for RCS Organizational Unit:

1. (Technical) The RCS shall be capable of inserting the spacecraft into circular orbit according to design spec XX.
2. (Schedule) The RCS shall be delivered by DD/DD/DDDD.
3. (Cost) The total cost for the RCS shall not exceed $$.

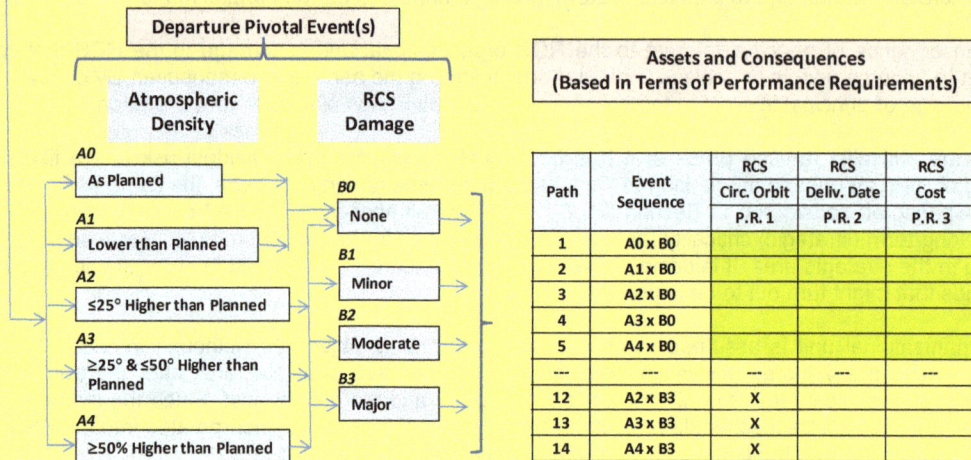

Departure Pivotal Event(s)

Atmospheric Density	RCS Damage
A0 **As Planned**	B0 **None**
A1 **Lower than Planned**	B1 **Minor**
A2 **≤25° Higher than Planned**	B2 **Moderate**
A3 **≥25° & ≤50° Higher than Planned**	B3 **Major**
A4 **≥50% Higher than Planned**	

Assets and Consequences (Based in Terms of Performance Requirements)

Path	Event Sequence	RCS Circ. Orbit P.R. 1	RCS Deliv. Date P.R. 2	RCS Cost P.R. 3
1	A0 x B0			
2	A1 x B0			
3	A2 x B0			
4	A3 x B0			
5	A4 x B0			
---	---	---	---	---
12	A2 x B3	X		
13	A3 x B3	X		
14	A4 x B3	X		

Figure 55. Expanded RSD for the Individual Risk Associated with Planet X Atmospheric Uncertainties from the Perspective of the RCS Organizational Unit

Based on this RSD, the probability of not achieving a circular orbit will be the summation over all pathways from 6 through 14 of the pathway probability times the consequence probability given the occurrence of the pathway. The quantification of this risk will be considered in Section 4.3.3.3.

4.3.3.1.1   Cross-Cutting Risks

The definition of cross-cutting risks given below is consistent with the definition provided in NPR 8000.4A.

Cross-Cutting Risks

A cross-cutting risk is an individual risk, with attributes and impacts found in multiple levels of the organization or in multiple organizations within the same level.

Generally speaking, the elements of a risk that make it cross-cutting are the condition and/or departure. It is important to determine whether or not an individual risk is cross-cutting because:

- Its total impact on performance risk is a function of its scope across organizational units.

- Decisions pertaining to its management are most effectively made at a level that encompasses its scope across organizational units.

The cross-cutting character of a given risk is best determined by an organizational unit at a level above the level at which that risk is first identified. Consequently, determination that an individual risk is cross-cutting depends on:

- Risk *identification* protocols that elicit information that supports cross-cutting risk determination

- Risk *analysis* protocols that systematically assess the cross-cutting potential of individual risks

- Risk *planning* protocols that elevate awareness of individual risks having the greatest potential to be cross-cutting, to the level in the organizational hierarchy where a determination of whether or not the risk is cross-cutting can most effectively be made

- Risk *communication* protocols that include knowledge management tools to communicate information pertaining to the cross-cutting potential of individual risks among affiliated organizational units.

Departure and asset taxonomies are useful for identifying cross-cutting risks. When standardized across organizational units, departure taxonomies can categorize individual risks into sets of cross-cutting risks affected by common departure events. Similarly, asset taxonomies can structure the postulation of reported individual risks onto other assets within an organizational unit's purview.

When an individual risk is cross-cutting and affects several organizational units, it is important that the relevant risk information be communicated to each unit. In this way, all the units affected by a cross-cutting risk can start from the same basis and produce consistent analyses that treat the risk as a common vulnerability. Risk scenario diagrams are an excellent way to ensure that this communication occurs, as will be demonstrated in the next subsection.

4.3.3.1.2   Incorporating RSDs from Other Organizational Units into One's Own

When several organizational units have risks that emanate from the same condition and depend, at least in part, on the same departure event, it is important for these organizational units to be aware of this commonality. With respect to the RSDs produced by the different organizations, the ultimate goal is for them to be consistent and unified in the areas where they are cross-cutting. In this way, the collection of RSDs produced within the project forms an integrated and consistent representation of how each cross-cutting risk affects each organizational unit.[28]

This concept is illustrated schematically in Figure 56. Organizational unit "Blue" identifies an individual risk and constructs an RSD consisting of a condition, several pivotal events, and several consequences. All of these elements first identified by "Blue" are denoted in blue. Organizational unit "Green," on the same level as "Blue," determines that it also has an individual risk with the same condition and some of the same pivotal events as "Blue," but its consequences are different from "Blue's," so some of the pivotal events leading to them are also different. The unique elements for "Green" are denoted in green. Organizational unit "Gray" resides above both units in the organizational structure, and its consequences are influenced by both "Blue" and "Green" who are suppliers to "Gray". Thus, there is an individual risk pertinent to "Gray" that derives from the same condition that "Blue" has identified and has some of the same pivotal events as "Blue" and "Green." Its consequences, however, are different from both "Blue's" and "Green's," and so "Gray" has some unique pivotal events that apply specifically to it. Its RSD contains some elements that appear in "Blue's" and "Green's" RSDs, but contains other elements that are unique to it. The unique elements for "Gray" are denoted in gray.

The yellow box following Figure 56 provides an example for how the RSD derived for Risk 1(b), RCS damage caused by unexpected atmospheric conditions, can be integrated into the RSD for Risk 1(a), planetary Pu contamination caused by unexpected atmospheric conditions. The former is constructed at the RCS level and the latter at the spacecraft level.

---

[28] An analogy is how DNA in human beings contains elements that are replicated across a large number of individuals. While the composite gene structure is specific to the individual, many of the building blocks that comprise it are identical.

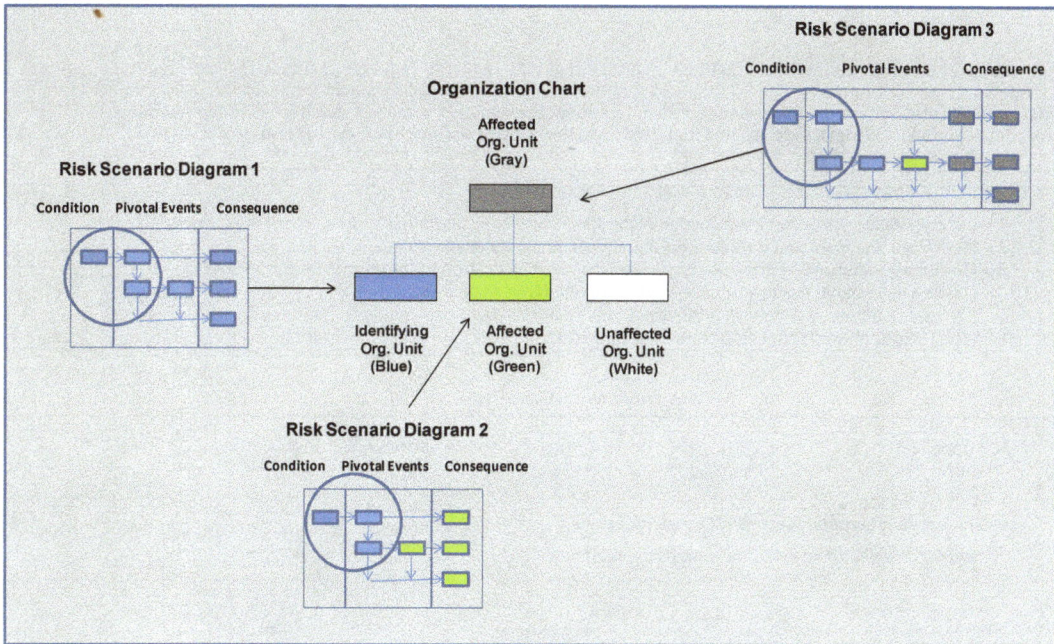

Figure 56. Interfacing of Risk Scenario Diagrams for Different Organizational Units

Planetary Science Mission Example: Integrated Risk Scenario Diagram for Planetary Contamination Caused by the Atmospheric Density being Higher than Expected

The spacecraft organizational unit has developed a rudimentary RSD for this individual risk that accounts for multiple pivotal events and consequences (see Figure 55). Since the strategic criticality of this individual risk issue is rank 3, which is very high, the unit determines that it needs to proceed with a relatively high level of detail in incorporating the individual risk into the aggregate performance risk models. However, it has also learned that the RCS unit has identified an individual risk that derives from the same condition (uncertain atmospheric density) and leads to departure events that may cause risk to the spacecraft unit's performance requirements. Therefore, the spacecraft organizational unit needs to ensure that its RSD incorporates the elements from the RCS RSD that affect both organizational units in common. In addition, it needs to ensure that its RSD clearly identifies the performance requirements being affect and is in the proper format.

The figure below shows the RSD that the spacecraft unit now develops. It contains pivotal events that are contained in the RCS unit's RSD (shaded in blue) as well as additional pivotal events that are relevant to its own performance requirements (shaded in gray). This RSD will serve as the conceptual basis for the model it will implement in its aggregate performance risk models.

Risk 1(a): Plutonium Contamination (Owned by Spacecraft Org Unit)

Given that [CONDITION: the state of knowledge of Planet X's atmosphere is limited; the fact that it is difficult to ascertain more information about Planet X's atmosphere from Earth; and the fact that the spacecraft contains radioactive material]

Performance Requirements for the Spacecraft Organizational Unit:

1. (Safety) There shall be no contamination of the planet surface by plutonium carried aboard the spacecraft
2. (Technical) The spacecraft shall successfully rendezvous with Planet X
3. (Technical) Video images of the planet surface shall be obtained in accordance with Design Spec XX.
4. (Technical) The total mass of the spacecraft shall not exceed XX pounds
5. (Schedule) The launch date shall be no later than DD/DD/DDDD.
6. (Cost) The total cost for the project shall not exceed $$.

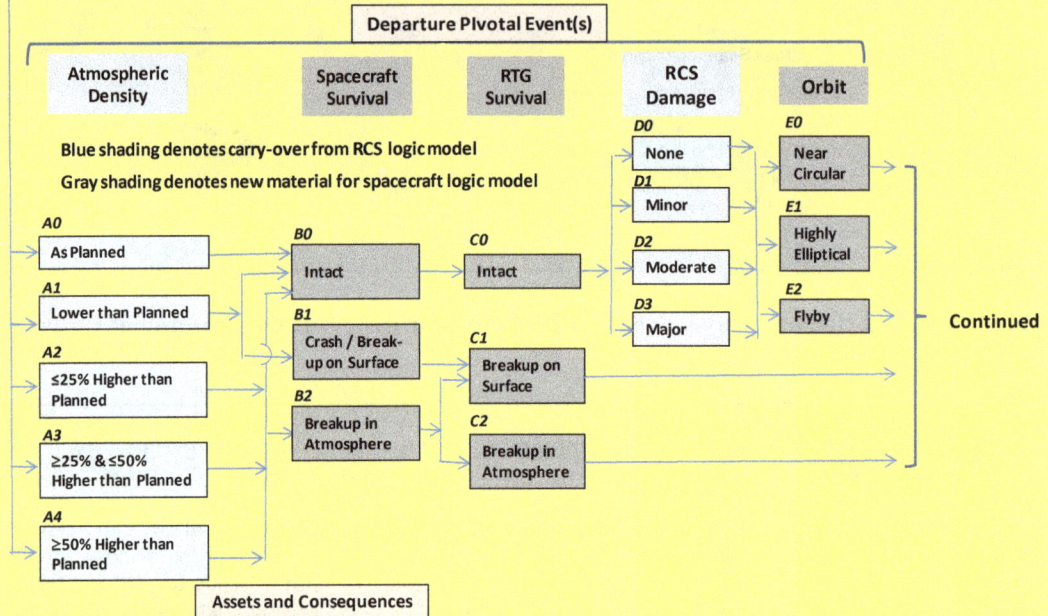

Departure Pivotal Event(s)

Atmospheric Density	Spacecraft Survival	RTG Survival	RCS Damage	Orbit

Blue shading denotes carry-over from RCS logic model

Gray shading denotes new material for spacecraft logic model

Atmospheric Density
- A0: As Planned
- A1: Lower than Planned
- A2: ≤25% Higher than Planned
- A3: ≥25% & ≤50% Higher than Planned
- A4: ≥50% Higher than Planned

Spacecraft Survival
- B0: Intact
- B1: Crash / Break-up on Surface
- B2: Breakup in Atmosphere

RTG Survival
- C0: Intact
- C1: Breakup on Surface
- C2: Breakup in Atmosphere

RCS Damage
- D0: None
- D1: Minor
- D2: Moderate
- D3: Major

Orbit
- E0: Near Circular
- E1: Highly Elliptical
- E2: Flyby

Continued

Assets and Consequences

Path	Event Sequence	S/C Contam-ination P.R. 1	S/C Rendez-vous P.R. 2	S/C Images P.R. 3
1	A0 x B0 x C0 x D0 x E0			
2	A1 x B0 x C0 x D0 x E0			
3	A2 x B0 x C0 x D0 x E0			
---	---	---	---	---
41	A0 x B0 x C0 x D0 x E2		X	X
42	A1 x B0 x C0 x D0 x E2		X	X
43	A2 x B0 x C0 x D0 x E2		X	X
---	---	---	---	---
64	A2 x B2 x C2	X	X	X
65	A3 x B2 x C2	X	X	X
66	A4 x B2 x C2	X	X	X

Continued

Expanded RSD for the Individual Risk Associated with Planet X Atmospheric Uncertainties from the Perspective of the Spacecraft Unit.

4.3.3.2 Updating the Performance Risk Models and Calculating Performance Risk

The process for integrating the individual risks into an aggregate model to evaluate performance risks is the same as has been discussed in Sections 3.2.1 and 3.2.2. The mechanics of the process were highlighted in Figure 23.

With respect to that figure, there are a few points that are pertinent to the Analyze step in CRM. Portions of the figure are copied here to facilitate the discussion of those points.

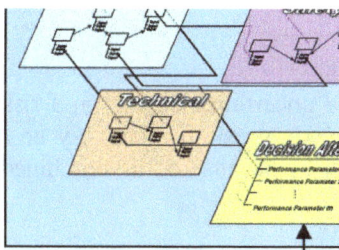

Existing Risk Analysis Framework & Models

With respect to risk analysis framework and models, it will be necessary to include the logic developed in the new RSDs for cases where the strategic criticality ranking of the individual risk is high. As discussed earlier within the corresponding RIDM Section 3.2.2.1, it will also be necessary to account for interactions or duplications between events in the new and existing scenarios.

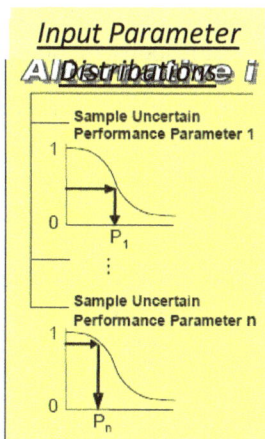

Input Parameter Distributions

With respect to input parameter distributions, some of the pivotal event probabilities will be obtained from models rather than from estimates obtained from experience. For example, determination of the probability that the spacecraft will burn up in the planetary atmosphere during the aerocapture maneuver may be based on models of how the trajectory is affected by atmospheric density and how the margin of safety for the heat shield is affected by the trajectory. Similarly, the probability of exceeding the launch window due to a software utility becoming nonfunctional may be based on models that relate the amount of slippage in the launch date to the timeframe during the project when the software is nonfunctional (i.e., the initial failure time and the duration of the failure). Thus, new performance parameters will be introduced into the models as needed to characterize the pivotal event probabilities. If the strategic criticality of the individual risk is low, it may be sufficient to use best estimates or, when appropriate, bounding estimates for these performance parameters even though there is uncertainty associated with them. If the criticality is high, however, it will be necessary to derive probability distributions for these performance parameters.

Performance Measure Values

#	Measure 1	...	Measure n
1	$PM_{1,1}$		$PM_{n,1}$
2	$PM_{1,2}$		$PM_{n,2}$
3	$PM_{1,3}$	...	$PM_{n,3}$
⋮	⋮		⋮
N	$PM_{1,N}$		$PM_{n,N}$

With respect to performance measure values obtained from the simulation using the newly developed models and parameter distributions, these values will have to be quantified for each decision point during the project and compared to the risk tolerances provided by the decision maker at each decision point.

4.3.3.3 Determining the Risk Drivers

As mentioned in the Section 4 introduction, risk drivers are those uncertain elements in a risk analysis (elaborated below) that contribute most to performance risk. A risk element may be a driver because of its significant contribution to an individual risk or because of its combined influence in a number of individual risks.

Risk drivers are determined after the performance risk models have been created and executed by performing sensitivity analyses within the performance risk calculations. The purpose of determining drivers is not to determine the urgency of individual risks or the level of detail to be applied to each individual risk, but rather to assist the development of planning response options.

Risk drivers may be of several kinds. A parameter in the model for determining the probability of a pivotal event may be a risk driver. Likewise, the pivotal event itself may be a risk driver if its probability is evaluated without the use of a parametric model. Likewise, the departure in the risk statement may be a risk driver if the departure is not expanded into pivotal events via a risk scenario diagram. Even when a departure is expanded into pivotal events and the pivotal event probabilities into parameter models, it is useful to define the risk driver in terms of all three elements: the parameter that drives the pivotal event probability, the pivotal event whose probability drives the departure probability, and the departure whose probability drives the performance risk. Doing so helps ensure that all appropriate mitigation options are considered and that overall mitigation approach adheres to the principles of a "defense-in-depth" strategy.[29]

The identification of risk drivers may sometimes need to be performed in two steps or more. The first step looks at each parameter, pivotal event, or departure individually to determine whether it is a driver by itself. If no drivers are identified by this process, then combinations of parameters, pivotal events, and/or departures are considered.

---

[29] Defense-in-depth refers to creating multiple independent and redundant layers of defense to compensate for potential failures so that no single layer, no matter how robust, is exclusively relied upon.

The following tasks outline one method for determining risk drivers. Other methods may be used if deemed more appropriate, but they should all be based on a sensitivity analysis approach to determining the relative importance of each parameter, event, or departure. These tasks are performed for each decision point in time to determine whether an individual parameter, pivotal event, and/or departure is a risk driver:

- From the estimated uncertainty range for each performance parameter, pivotal event probability, and/or departure probability, specify what would qualify as an optimistic value (e.g., a 95% confidence value).

- Evaluate what the value of the aggregate performance risk would be if one performance parameter, pivotal event probability, and/or departure event probability were placed at its optimistic value while all others were left unchanged.[30]

- Compare this value of the performance risk with the value obtained prior to making the single change described above.

- Designate the parameter, pivotal event, or departure as a driver if the single change causes the performance risk to cross one or more tolerability thresholds (i.e., yellow to green, red to yellow, or red to green).

In the event that no individual drivers are identified for a performance risk that is yellow or red, then perform the following tasks:

- From the above analysis, determine the top two parameters, pivotal events, and/or departures that have the largest effect on the performance risk. Set them at their optimistic values while the others are unchanged, and repeat the tasks above. If the performance risk crosses one or more tolerability thresholds, designate the combination of the two parameters as a risk driver.

- If necessary, look at higher order combinations (i.e., three or more) until the performance risk crosses a tolerability threshold.

The yellow box below illustrates how risk drivers may be derived for the performance risk associated with failure to achieve a circular orbit after the aerocapture maneuver (Performance Requirement 1 in Figure 55).

---

[30] Some model parameters, pivotal event probabilities, and departure probabilities could be random variables having uncertainty distributions assigned to them as in Figure 23. Others could have assessed values obtained from best estimate analysis. In the sensitivity analysis, the parameter, pivotal event, or departure being investigated would have its uncertainty distribution or best estimate value replaced by the estimated optimistic value.

This example concerns the individual risk whose RSD is shown in Figure 55. Several models are used to predict the magnitudes of relevant performance parameters and associated probabilities. The results are as follows:

A 6-degree-of-freedom trajectory model predicts that:

- 8 of 12 thrusters in the RCS must be working in order for the spacecraft to achieve circular orbit given that the atmospheric density is as planned
- 9 of 12 must be working given the atmospheric density is lower than planned
- 10 of 12 must work given the density is 25% higher than planned
- 12 of 12 must work given the density is 50% higher than planned

The uncertainty in these predictions is estimated to be ±1 thruster for all four cases.

A structural response model for the spacecraft predicts that:

- The 2 most windward thrusters will fail if the density is 25% higher than planned
- The 4 most windward thrusters will fail if the density is 50% higher than planned.

The associated uncertainty is ±2 thrusters for both cases.

A statistical analysis of the available data on the planetary atmospheric shows that on any particular day at any particular location:

- There is a 10% likelihood that the density is at least 25% higher than expected
- There is a 5% likelihood that the density is at least 50% higher than expected.

From this information, the aggregate performance risk model predicts that the total probability for failure to achieve a circular orbit is 0.005 at the time of SIR (the most critical decision point for this risk). This model includes the individual risk of RCS damage during the aerocapture maneuver together with all other individual risks that contribute to the likelihood of not achieving a circular orbit.

A set of sensitivity calculations is performed using the aggregate risk model to determine whether any of these uncertainties in the risk calculation are risk drivers. The results are shown in the table below:

Sensitivity Case	Description	Calculated Probability of Not Achieving Circular Orbit	Tolerability Thresholds	Risk Driver?
0	Base Case	0.005	0.001 for intolerable to marginal	No
1	2 fewer thrusters (than in the base case) failed because of loading during aerocapture	0.003		No
2	1 fewer thruster (than in the base case) needed to achieve circular orbit after aerocapture	0.003	0.002 for marginal to tolerable	No
3	Combination Case 1 and 2	0.001		Yes

From these results, none of the elements by themselves are risk drivers. However, the combination of two elements is a risk driver:

The table below illustrates how the identified risk driver might appear in a list of risk drivers. In this case, since it is defined down to parameter level, the characterization of the driver includes the departure, the pivotal event, and the performance parameters.

Risk Driver No.	Affected Org. Unit	Risk Driver Characterization		
		Departure	Pivot Event	Parameter *
1	RCS	Probability of abnormal aerocapture trajectory precluding circular orbit	Probability of RCS damage during aerocapture precluding circular orbit	Uncertainty in the number of thrusters that fail due to atmospheric density ≥ 25% higher than planned
				Uncertainty in the number of thrusters needed to achieve a circular orbit following aerocapture due to atmospheric density ≥ 25% higher than planned
2	RCS	TBD	TBD	TBD

* Note: These are uncertainties in parameters that are input to the aggregate performance risk calculation. However, they are also outputs from deterministic calculations using trajectory and structural response models. Thus, they could also be referred to as "Model Uncertainty Drivers".

4.4 The CRM Plan Step

The CRM *Plan* step addresses what, if any, action should be taken to address an activity's performance risk. As discussed in Section 4.3, there are two primary pathways into the *Plan* step:

- The pathway leading from the "tactical" quick-look risk analysis of individual risks, which can identify a critical individual risk that needs a rapid response due to a short window of opportunity for managing it; and

- The pathway leading from the "strategic" in-depth risk analysis of the entire activity, using the current risk model to identify the risk drivers that are the greatest contributors to performance risk.

These two pathways are illustrated in Figure 57 and Figure 58, respectively.

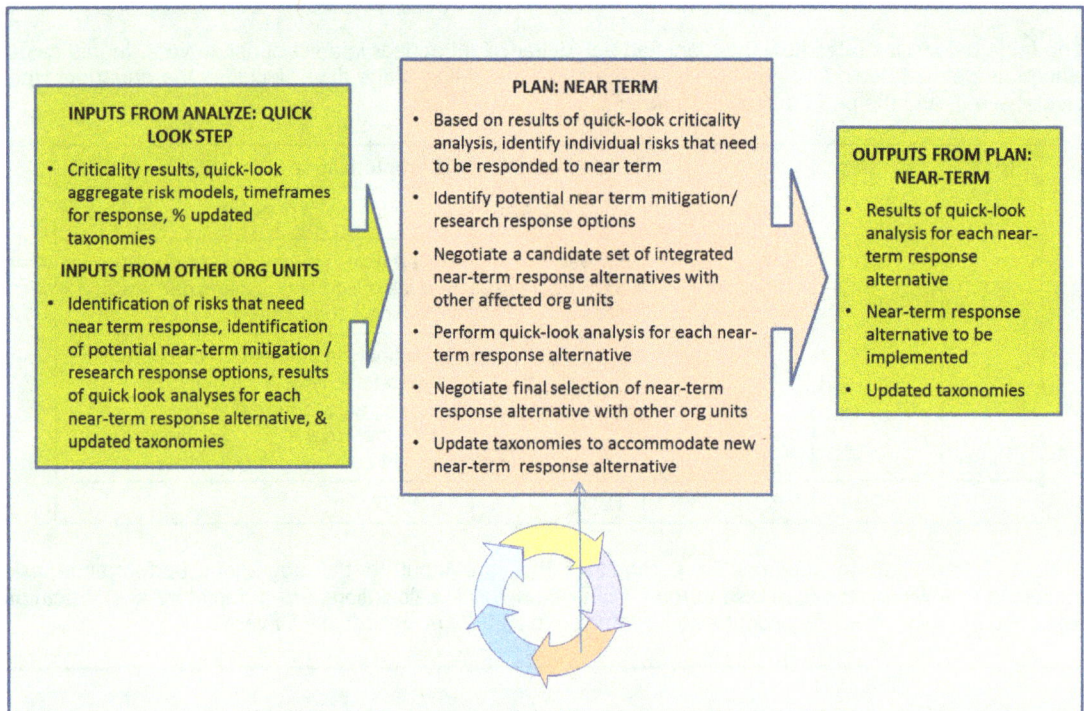

INPUTS FROM ANALYZE: QUICK LOOK STEP

- Criticality results, quick-look aggregate risk models, timeframes for response, % updated taxonomies

INPUTS FROM OTHER ORG UNITS

- Identification of risks that need near term response, identification of potential near-term mitigation / research response options, results of quick look analyses for each near-term response alternative, & updated taxonomies

PLAN: NEAR TERM

- Based on results of quick-look criticality analysis, identify individual risks that need to be responded to near term
- Identify potential near term mitigation/ research response options
- Negotiate a candidate set of integrated near-term response alternatives with other affected org units
- Perform quick-look analysis for each near-term response alternative
- Negotiate final selection of near-term response alternative with other org units
- Update taxonomies to accommodate new near-term response alternative

OUTPUTS FROM PLAN: NEAR-TERM

- Results of quick-look analysis for each near-term response alternative
- Near-term response alternative to be implemented
- Updated taxonomies

Figure 57. The CRM Plan Step (Tactical Response)

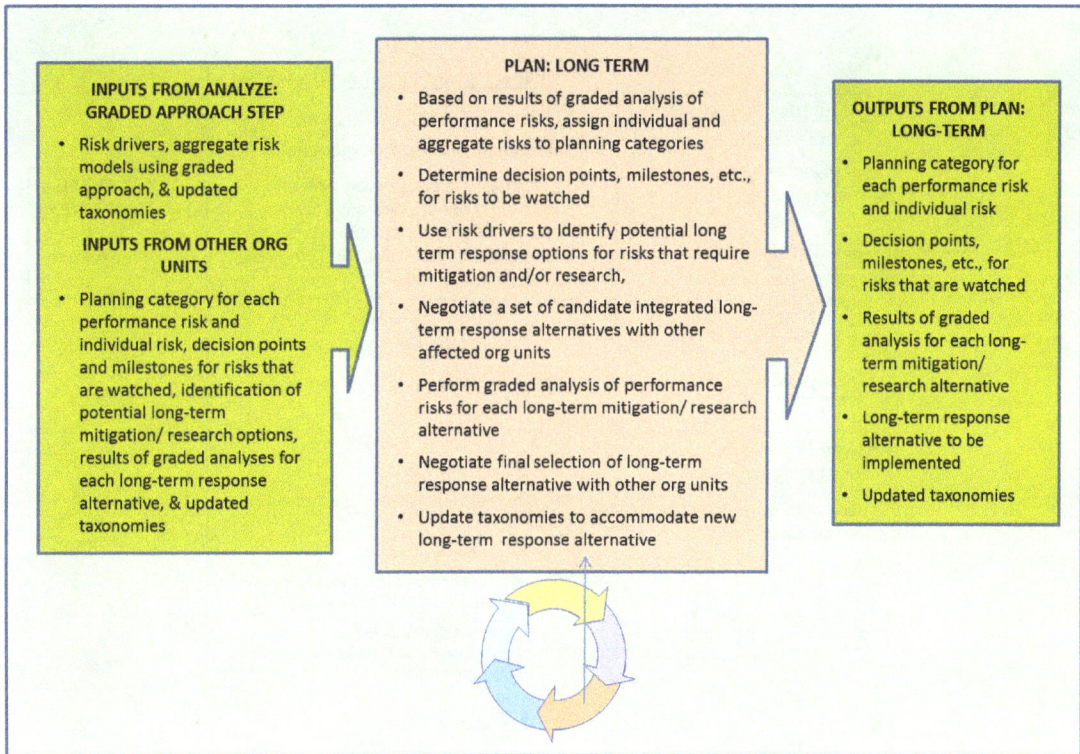

Figure 58. The CRM Plan Step (Strategic Response)

For both the tactical and strategic dimensions of CRM *Plan*, the same four basic tasks occur, as shown in Figure 59. These tasks are 1) generate a set of candidate risk response alternatives; 2) conduct a risk analysis of each alternative; 3) deliberate and select an alternative for implementation, and 4) implement the selected alternative. The main differences between the two dimensions of *Plan* are the level of urgency and the level of analysis that has been done to support risk response decision making. In the tactical *Plan*, the presumption is that a risk response is needed quickly, leaving insufficient time for a thorough analysis of the individual risk and integration of the risk into the risk model. Instead, decision making would rely more heavily on qualitative engineering judgment to determine the best response, at least in the near term.

However, as shown in the CRM process flow diagram (Figure 40 in Section 4), all individual risks are subject to in-depth risk analysis, regardless of whether or not they are also subject to "tactical response" planning. This is necessary in order to keep the risk model current and reflective of all identified sources of performance risk – a necessary precondition of the strategic *Plan* step, which is done in the context of an integrated risk model that reflects the full spectrum of risk that has been identified to date. Indeed, in order for the risk model to reflect the current reality, it is also necessary that it include all risk responses that have been implemented to date, whether from strategic or tactical planning.

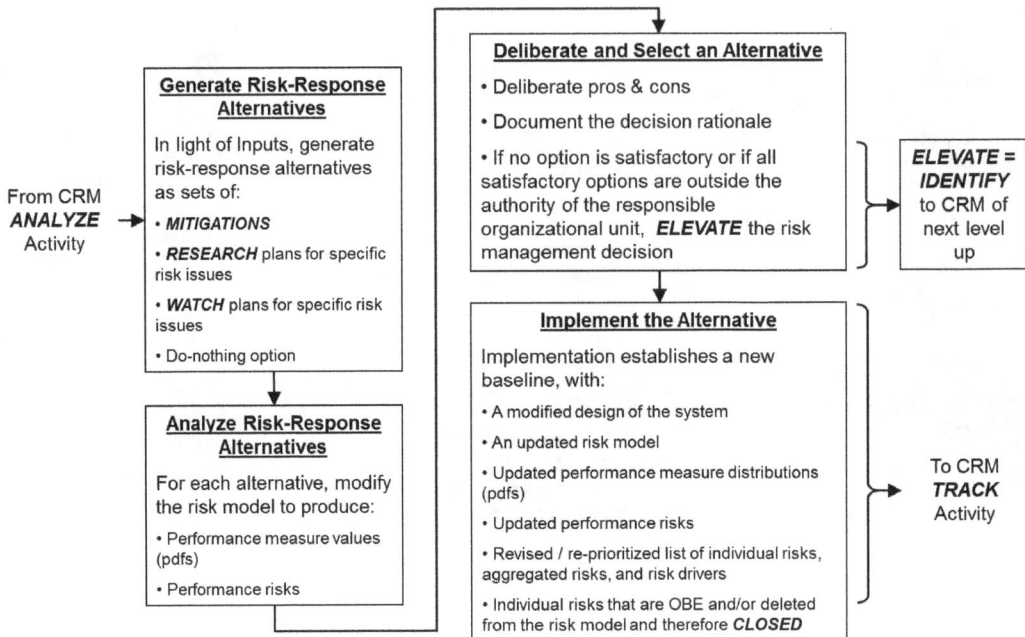

Figure 59. CRM Plan Tasks

As is the case with all CRM steps, the CRM *Plan* step is executed throughout the life of the program / project. However, the timeframe of its execution differs between the tactical and strategic steps:

- Tactical CRM planning is triggered whenever an individual risk is forwarded for a rapid risk response due to its analyzed criticality and urgency. To properly execute tactical CRM planning, a unit's risk management function must have the capacity to act quickly when called upon to determine how best to respond to these urgent risks.

- Strategic CRM planning is conducted if, after an identified risk has been fully integrated into the risk model (including any implemented tactical risk responses), the analyzed performance risk is out of tolerance. When this is the case, strategic CRM planning brings the full risk analysis capability to bear on the generation and analysis of the set of candidate risk response alternatives, so that the selection of an alternative for implementation is fully risk informed.

The CRM *Plan* step is a type of risk-informed decision making, in which the decision of how best to address the risk posture of an activity is informed via risk analysis of the set of candidate risk response alternatives. However, unlike the RIDM process in Section 3 of this Handbook, CRM planning is done in the context of defined performance requirements; there is no need to conduct those RIDM activities related to requirements setting, such as the establishment of performance commitments.

4.4.1 Generating Risk Response Alternatives

Experience has shown that risk responses can be multidimensional, involving a number of discrete responses that act together to reduce performance risk.[31] In the terminology of this Handbook, risk responses may consist of a number of individually defined risk response options, each of which is of a particular risk disposition type specified in NPR 8000.4A. The process of generating a set of candidate risk response alternatives consists of first generating a set of candidate risk response options. The risk response alternatives are then defined as different combinations of these options, including alternatives consisting of a single option. Figure 60 illustrates the concept.

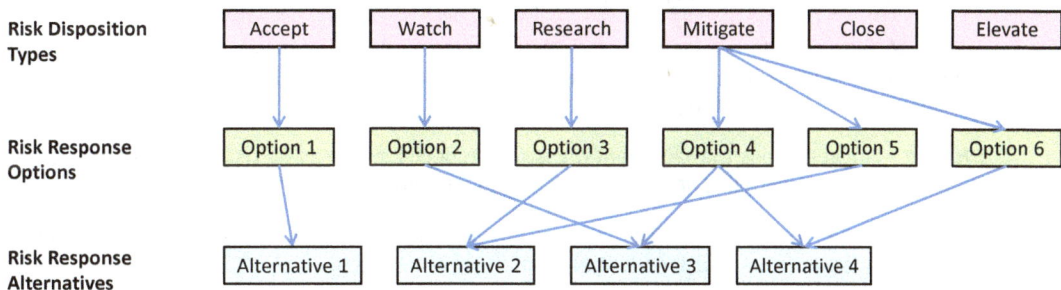

Figure 60. Relationship between Risk Response Options and Risk Response Alternatives

4.4.1.1 Generating Risk Response Options

Risk response options are the individual responses from which candidate risk response alternatives are generated. Each risk response option pertains to one of the following risk disposition types specified in NPR 8000.4A: Accept, Mitigate, Close, Watch, Research, and Elevate.

---

[31] This is the case when, for example, foam shedding is addressed by two risk response options simultaneously: a mitigation consisting of the reshaping of foam structures to reduce aerodynamic stresses; and a research activity to better understand the fundamental physics behind foam separation under launch conditions.

The following bullets provide guidance on when to apply each disposition:

- **Accept** – A risk response of *Accept* indicates that no risk management action needs be taken, given the current analyzed performance risk. This is typically because the performance risks associated with the performance requirements are all within tolerable levels, reflecting an activity that is on track to accomplish its objectives within established risk tolerances. Within the risk model, this means that at the time of the risk analysis, none of the identified risk drivers are of sufficient magnitude to create intolerable performance risk.

  However, a risk response of *Accept* does not mean that no risk management action relating to the existing risk drivers will be needed in the future. As the activity proceeds, additional conditions and departures may be identified that compound the effects of existing risk drivers in a manner that produces intolerable performance risk. In such cases, risk drivers that previously did not warrant a risk response might now be the most attractive targets for reducing performance risk.

  As the "no action" risk response option, the *Accept* option does not combine with other options when generating candidate risk response alternatives. An option of *Accept* applies to the entirety of the activity's risk posture and is superseded by any other risk response.

  A risk response of *Accept* must be documented by the organizational unit, including the assumptions and conditions on which it is based.

- **Mitigate** – A risk response option within *Mitigate* is the taking of positive action to address the activity's performance risk. This is typically because the performance risk of one or more performance requirements is outside tolerable bounds. However, it is important to allow for the possibility that mitigation may also be employed simply because an opportunity exists to reduce performance risk even when it is within bounds. Mitigation options typically address one or more risk drivers, and are focused on improving the performance risk where it is most in need of improvement, without producing too large a collateral increase of performance risk in other areas (e.g., cost and schedule). Because mitigation options can address one or more risk drivers, and because a single risk driver can be present in a number of individual risks, a single mitigation option can potentially be responsive to a substantial number of individual risks.

Mitigation can be classified into two broad categories: departure prevention and consequence reduction. Departure prevention refers to those risk response options that, if successfully deployed, prevent or reduce the likelihood of the departure event, and therefore the likelihood of the performance shortfalls associated with that event. Consequence reduction refers to those risk response options that, if successfully deployed, reduce the severity of the consequence produced by the departure, and therefore, of the magnitude of the associated performance shortfalls. Departure prevention and consequence reduction are illustrated graphically in Figure 61.

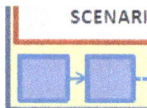

Figure 61. Departure Prevention and Consequence Reduction

A mitigation option can affect the derived requirements that flow down to organizational units at the next lower level of the NASA hierarchy. When this is the case, implementation of the mitigation option includes the negotiation of a rebaselined set of derived requirements among the affected units. In situations where risk drivers have been elevated from lower levels in the organizational hierarchy, it is not unexpected that mitigation would involve changes to the derived requirements, given that the rationale for elevation is the inability of the lower level to manage the risk to the very same requirements within their scope of authority and capability. However, it may also be the case that an organizational unit, when managing their own internally identified risk, identifies an attractive mitigation option that entails rebaselining of derived requirements. In either situation, all affected organizations should participate in defining the mitigation option(s).

In more extreme cases it might be advantageous for an organizational unit to consider mitigation options that go beyond the scope of the design solution chosen during RIDM. When this is the case, the RIDM process is re-executed, using current conditions, to select a new alternative that may or may not be within the scope of the existing design solution. Candidate alternatives would typically include the contending alternatives from the original RIDM activity, but might also include previously discounted alternatives that are now attractive, or other alternatives not previously considered but which, due to changed conditions, are now attractive. Re-execution of RIDM generally produces entirely new sets of derived requirements flowing down to lower level units in the NASA hierarchy, and the costs associated with such a major shift must be factored into the risk analysis of alternatives.

Mitigation plans are documented, including the appropriate parameters that will be tracked to determine the effectiveness of the mitigation.

- **Watch** – A risk response of *Watch* identifies one or more risk drivers that will be monitored according to a documented set of tracking requirements that include, at a minimum, the specific parameters to be watched and a monitoring schedule according to which the parameters will be observed. Additionally, depending on the circumstances, the watch plan may include contingency plans or other types of deferred decisions that will be invoked conditional on the results of the monitoring activity.

  Watching entails the periodic updating of the risk analysis with current values of the watched parameters, according to the monitoring schedule. Because the activity's risk is analyzed using a single, integrated risk model, it may be efficacious to coordinate, as much as practicable, the monitoring schedules of the watched parameters, so that the risk analysis is updated and evaluated in a "batch" fashion.

- **Research** – A risk response option of *Research* applies to one or more risk drivers whose uncertainties are large enough that they interfere with robust risk management decision making. The *Research* option seeks to reduce uncertainty concerning some aspect of a risk driver by actively generating additional information about it. It entails the development of a research plan that identifies the subject to be researched, the specific parameters about which information is expected to be generated, and a research schedule including timeframes for results (and integration of the results into the risk model). Additionally, like the *Watch* option, the Research option may include contingency plans or other types of deferred decisions that will be invoked depending on the results of the research.

- **Elevate** – A risk response of *Elevate* transfers the management of a performance risk to the organizational unit at the next higher level. Elevation occurs when no satisfactory combination of *Mitigate*, *Watch*, and *Research* options can be found that return the risk to tolerable levels. The Elevate option recognizes that the inability to manage performance risk at one level of the NASA hierarchy directly impacts the performance risk at the next higher level. As discussed in Section 4.2.2, *Sources of Risk Identification*, elevation is a pathway into the CRM *Identify* step of the organizational unit at the higher level.

It is expected that the *Elevate* option will typically be combined with a *Watch* option that monitors the status of any risk drivers that may be associated with the unmanageable performance risk. In addition, the Elevate option may be combined with other options that address some fraction of the risk, though not enough to bring it to within tolerable levels. An *Elevate* option also entails coordination of the risk modeling activities of the organizational unit that is elevating the management of risk, and the unit to which the management of risk has been elevated. It is expected that the analysis of risk response alternatives will involve a coordinated risk analysis effort by both levels. As a practical matter, when elevating, the elevating unit should propose alternatives that it considers attractive but which exceeds the scope of its authority to implement.

- **Close** – A risk response of *Close* applies to individual risks whose risk drivers no longer exist or are no longer cost-effective to watch. This can occur when their probability has been reduced below a defined level of insignificance; the consequence potential has been reduced below a defined level of insignificance; or the event has occurred, thus becoming a problem rather than a risk management issue (and is tracked as such). Closing an individual risk indicates not only that it is currently not a significant contributor to performance risk, but that there is no expectation that it will be a significant contributor to performance risk in the future.

The yellow boxes below provide examples of how candidate mitigation/research options might be derived from information about risk drivers.

Planetary Science Mission Example: Identification of Candidate Mitigation/ Research Options Based on Risk Driver Characterization

As a result of the graded analysis of the performance risks, suppose that three risk drivers were identified. One task in planning is to identify candidate mitigation and/or research actions that could address these drivers. The information presented below characterizes each of the risk drivers and the corresponding candidate mitigation and/or research actions.

Risk Driver: 1

Affected Organizational Unit: RCS

Affected Individual Risk:

CONDITION: the state of knowledge of Planet X's atmosphere is limited; and the RCSs fielded to date have not needed to operate in such harsh (hyperbolic entry) environments

DEPARTURE: unanticipated atmospheric characteristics

ASSET: the exposed RCS components

COSEQUENCE: damage to the RCS system making it unable to perform orbital maneuvers necessary to achieve circular orbit

Pivotal Event Driver: Probability of RCS damage during aerocapture leading to inability to achieve circular orbit

Parameter Drivers:

1. Uncertainty in the number of thrusters that fail due to atmospheric density $\geq 25\%$ higher than planned

2. Uncertainty in the number of thrusters needed to achieve a circular orbit following aerocapture due to atmospheric density $\geq 25\%$ higher than planned

Candidate Mitigation / Research Option #1:

Description: Add thrusters to the RCS and locate them symmetrically with respect to the windward streamline but away from windward.

Potential New Risks: Additional mass might exceed the mass budget. May adversely affect spacecraft center of gravity. Redesign may affect delivery date and cost for both RCS and spacecraft.

Candidate Mitigation / Research Option #2:

Description: Increase the structural strength of the existing RCS thrusters on the windward side.

Intent: Circumvent parameter drivers.

Potential New Risks: Same as for Option #1. Additionally, the increased size of some thrusters might adversely affect the flow field causing shock interactions.

Risk Driver: 2

Affected Organizational Unit: Spacecraft

Affected Individual Risk:

CONDITION: the state of knowledge of Planet X's atmosphere is limited

DEPARTURE: unanticipated atmospheric characteristics

ASSET: the spacecraft

CONSEQUENCE: spacecraft breakup and radioactive contamination of Planet X

Pivotal Event Driver: Probability of the atmospheric density being high enough to cause the spacecraft to burn up

Parameter Drivers:

1. Uncertainty in the boundary layer transition time
2. Uncertainty in the turbulent heat flux

Candidate Mitigation / Research Option #1:

Description: Incorporate an impact probe that will be fired into the atmosphere prior to the aerocapture maneuver to obtain real time data on the atmospheric density and feed that information to the trajectory control system.

Potential New Risks: Additional mass might exceed the mass budget. May adversely affect spacecraft center of gravity. Launch date and project cost might be adversely affected by the need to develop new technology for the impact probe, software requirements, redesign of the spacecraft, and procurement of new hardware.

Candidate Mitigation/Research Option #2:

Description: Perform ground testing in Planet-X-type environments to reduce uncertainties in the modeling of the boundary layer transition time and the turbulent heat flux. Use results to redesign thickness of heat shield.

Potential New Risks: Testing and heat shield redesign will affect project schedule and cost. Additional mass of heat shield may necessitate additional structural thickness, which together may exceed the mass budget. No guarantee that results of testing will be conclusive or that risk will be reduced as a result of it.

Risk Driver: 3

Affected Organizational Unit: Electrical Power / RTG, subsequently elevated to project level

Affected Individual Risk:

CONDITION: plutonium is not presently being produced in this country and Congress has not approved funding for production and the supply is decreasing

DEPARTURE: the price of plutonium may drastically increase

ASSET: electrical power / RTG

CONSEQUENCE: cost exceeds funding

Pivotal Event Drivers:

1. Probability that Congress will not authorize funding for plutonium processing in the US
2. Probability that Russia will significantly raise their price of plutonium after the current contract expires

Parameter Drivers: None identified

Transfer of Ownership: Elevated to project level because of insufficient funding and/or authority at the lower level to resolve the risk

Candidate Mitigation/Research Option #1:

Description: Replace some of the heritage RTG power with the new Advanced Stirling Radioisotope Generators (ASRGs), which require less plutonium.

Potential New Risks: The new ASRGs have not been tested for long-term reliability and they may fail before the mission is completed. This could result in premature loss of power, which could lead to loss of science, loss of mission, and/or planetary contamination depending on when power becomes depleted.

Candidate Mitigation/Research Option #2:

Description: Allocate additional funds to purchase plutonium by reducing costs for new technology development. The means for doing this is to scrap development of new high-tech atmospheric sensors and use heritage equipment.

Potential New Risks: The heritage sensors may not achieve the required degree of resolution needed to fully characterize the atmosphere, since they were designed for a different set of atmospheric conditions.

Candidate Mitigation/Research Option #3:

Description: Replace the nuclear power sources with solar arrays, which would remove the issue of Pu contamination and eliminate the need for White House launch approval.

Potential New Risks: It may not be possible to achieve the power requirements for the extended life of the mission without exceeding weight limitations.

4.4.1.1.1 Communicating Risk Response Options

Risk response option generation should involve all affected organizational units. Communication aids and protocols should be established that inform other potentially affected units of each unit's CRM *Plan* activities, so that synergies and cross-cutting risks can be identified early in the planning process. A Risk Driver List is one such communication aid. Another aid [being defined here] is the "Risk Response Plan," which is a document that is developed over the course of the *Plan* step that identifies the risk drivers against which risk management has decided to act. The Risk Response Plan develops as risk management personnel compile risk response options (e.g., from the narrative descriptions) and propose additional options. At this point, the Risk Response Plan is shared with other potentially affected organizational units, that in turn share their Risk Response Plans. This enables multiple organizational units to work cooperatively to address the risk drivers, as a function of:

- The cross-cutting nature of the risk drivers

- The impact that the risk drivers have on performance risk

- The capacity of proposed risk response options to reduce the risk across multiple units.

4.4.1.1.2 Pruning Risk Response Options

Generation of candidate mitigation/research actions typically begins with qualitative brainstorming among knowledgeable personnel, in order to generate a broad initial set of options. This set is then pruned as it becomes evident that some options are either infeasible (i.e., they are incapable of being implemented within the requirements) or categorically inferior to other options. At this point in the process, simple engineering analysis may be appropriate to gain a level of quantitative understanding of the cost and effectiveness of each option. The risk analysis provides an analytical basis for testing and downselecting proposed risk response options.

In general, Step 2 of the RIDM process, Compile Feasible Alternatives (Section 3.1.2) is applicable to the generation of risk response options. However, it is not expected that each risk response option is capable of achieving the desired risk management outcome (e.g., performance risk reduction to tolerable levels) on its own. Instead, the risk response option set represents the pool from which subsets are proposed for implementation, based on synergies among options and the ability of the subsets to collectively produce the intended result.

4.4.1.1.3 Special Considerations for *Elevate* and *Close*

The risk response option types of *Elevate* and *Close* present unique considerations:

- A risk response of *Elevate* should only be made in response to an inability of the organizational unit to effectively manage performance risk at its level in the NASA hierarchy. As such, *Elevate* should not be proposed as an initial option. Instead, it should be reserved for situations in which the available risk response options (and alternatives) have been analyzed and shown to be inadequate. Only then should Elevate be proposed, in combination with other options, or as a stand-alone option.

- A risk response of *Close* does not affect performance risk. It is mainly a bookkeeping device to remove from consideration those individual risks and risk drivers that no longer warrant risk management attention. As such, *Close* options can be proposed without regard for other options that are explicitly intended to address performance risk. As a practical matter, it may be most efficient to *Close* all appropriate individual risks and risk drivers at the beginning of each execution of the strategic CRM *Plan* step, thereby reducing the number of issues against which planning takes place.

4.4.1.2 Combining Risk Response Options to Produce a Set of Candidate Risk Response Alternatives

Once a set of risk response options has been defined, a set of candidate risk response alternatives can be produced, where each risk response alternative consists of a combination of risk response options that can potentially work together to produce an adequate response to the current risk posture. Theoretically, for N risk response options it is possible to define 2^N candidate risk response alternatives (including the no-action alternative). However, in practice, the candidate risk response alternatives can be constrained to a reasonable number by downselecting attractive alternatives that:

- Address the performance risk of multiple organizational units

- Address all (or most) of the performance requirements whose performance risk is outside tolerable levels

- Introduce less performance risk in other requirements areas (e.g., cost, schedule) in order to achieve the intended risk reduction.

Risk response alternatives can be composed of risk response options of different types (e.g., *Mitigate* and *Watch*, *Research* and *Elevate*). Figure 62 illustrates a tabular method of identifying and downselecting candidate risk response alternatives for further consideration.

	Risk Response Option 1	Risk Response Option 2	Risk Response Option 3	Risk Response Option 4	Risk Response Option 5	Risk Response Option 6	Risk Response Option 7	Risk Response Option 8	Risk Response Option 9	Risk Response Option 10
Risk Response Alternative 1	X									X
Risk Response Alternative 2	X					X				
Risk Response Alternative 3	X			X						
Risk Response Alternative 4	X				X					
Risk Response Alternative 5								X		
Risk Response Alternative 6			X	X		X				
Risk Response Alternative 7					X					
Risk Response Alternative 8					X					
Risk Response Alternative 9				X	X		X			
Risk Response Alternative 10	X		X	X		X				

Figure 62. Notional Risk Response Matrix

4.4.2   Risk Analysis of Risk Response Alternatives

Once a set of risk response alternatives have been defined, risk analysis of the alternatives is conducted to risk-inform the selection of an alternative for implementation.

4.4.2.1 Integrating the Candidate Risk Response Alternatives into the Risk Analysis

For both tactical and strategic risk response planning the risk analysis is conducted by integrating each risk response alternative into the risk model and quantifying the resulting performance risk. The difference is one of detail; the level of rigor used to model the alternatives for tactical risk response planning is expected to mirror that used during the quick-look analysis to characterize the individual risk in the risk model. Point estimate parameter valuation, subjectively determined by relevant subject matter experts, is generally considered sufficient for tactical planning, where time may be of the essence. The use of simplified models represents a compromise between the desire for robust, risk-informed decision making and the need for action within a potentially short window of opportunity. (It is expected that the risk analysis of the selected alternative will subsequently be revised during the next detailed *Analysis* activity, to raise the overall level of rigor of the new baseline risk model to the appropriate standard.) For strategic planning, the level of rigor used to model the alternatives should be consistent with the existing standards of the risk model as it relates to the level of detail of the RSDs, the pedigree of the data used for quantification, and the treatment of uncertainty.

The next yellow box illustrates how a set of mitigation options (i.e., a mitigation alternative) may be incorporated into the RSD for Risk 1(b) in Appendix G: Planetary Pu contamination caused by unexpected atmospheric density. It further discusses how the probabilities of the pivotal events may be affected by the mitigation alternative.

Planetary Science Mission Example: Integrated Risk Scenario Diagram for Planetary Contamination Caused by the Atmospheric Density being Higher than Expected Including Mitigation Alternative 1

As a result of considering the various mitigation options that were identified from the risk drivers, the project in combination with its lower level organizational units has determined that three combinations of options will each be considered and evaluated as mitigation alternatives. The first of these includes the following actions:

- Strengthen the four most windward thrusters in the RCS to protect against higher than expected aerodynamic forces during the aerocapture maneuver

- Incorporate an impact probe to obtain real-time measurements of atmospheric density that can be fed back to the trajectory control software

- Redesign the mass distribution to ensure that the center of gravity is properly located

- Replace some of the RTG power with ASRG power to reduce the cost of purchasing plutonium

The spacecraft organizational unit has produced three amended versions of its baseline RSD to incorporate each mitigation alternative. The RSD corresponding to the first mitigation alternative is shown in the illustration below. Two pivotal events have been added (and color-coded as orange in the figure to indicate that they have been introduced as a result of the proposed mitigation alternative). The first considers whether the ASRGs are able to operate long enough to complete the mission. If not, then the minimum consequence in the technical domain will be a partial loss of video imaging due to the loss of electrical power while the spacecraft is in orbit. In the extreme case where the power produced by the ASRGs is lost prior to the rendezvous with the planet, the technical consequence is loss of mission. For all pathways in the RSD, there is also a possibility that there will problems meeting the requirements related to spacecraft mass, launch date, and project cost. These are inherent risks associated with the fact that the mitigation alternative involves mass increases, schedule delays, and additional costs, any of which could contribute to an intolerable performance risk.

In addition to changes in the RSD, the incorporation of the mitigation alternative may profoundly affect the probabilities of the pivotal events that were present in the baseline RSD. For example, if the mitigation alternative were successful, it would greatly reduce the probability of the spacecraft burning up in the atmosphere, the probability of it crashing into the planet, the probability of the RCS being damaged, and given that everything survives, the probability of not achieving the desired circular orbit.

Risk 1(a): Plutonium Contamination (Owned by Spacecraft Org Unit)

Given that [CONDITION: the state of knowledge of Planet X's atmosphere is limited; the fact that it is difficult to ascertain more information about Planet X's atmosphere from Earth; and the fact that the spacecraft contains radioactive material]

and that [MITIGATION PLAN: the structural strength of the RCS thrusters on the windward side will be increased; an impact probe will be fired into the atmosphere prior to the aerocapture maneuver; and some of the RTGs will be replaced with ASRGs]

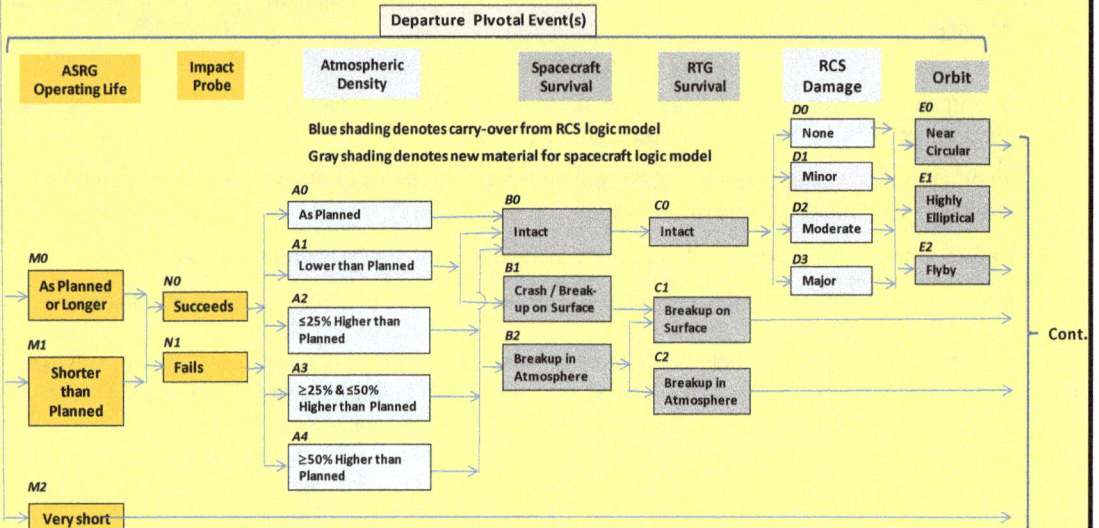

Departure Pivotal Event(s)

ASRG Operating Life	Impact Probe	Atmospheric Density	Spacecraft Survival	RTG Survival	RCS Damage	Orbit

Blue shading denotes carry-over from RCS logic model

Gray shading denotes new material for spacecraft logic model

D0 None — E0 Near Circular

D1 Minor

A0 As Planned — B0 Intact — C0 Intact — D2 Moderate — E1 Highly Elliptical

A1 Lower than Planned — D3 Major — E2 Flyby

M0 As Planned or Longer — N0 Succeeds — A2 ≤25% Higher than Planned — B1 Crash / Break-up on Surface — C1 Breakup on Surface

M1 Shorter than Planned — N1 Fails — A3 ≥25% & ≤50% Higher than Planned — B2 Breakup in Atmosphere — C2 Breakup in Atmosphere

A4 ≥50% Higher than Planned

Cont.

M2 Very short

Assets and Consequences

Performance Requirements for the Spacecraft Organizational Unit:

1. (Safety) There shall be no contamination of the planet surface by plutonium carried aboard the spacecraft
2. (Technical) The spacecraft shall successfully rendezvous with Planet X
3. (Technical) Video images of the planet surface shall be obtained in accordance with Design Spec XX.
4. (Technical) The total mass of the spacecraft shall not exceed XX pounds
5. (Schedule) The launch date shall be no later than DD/DD/DDDD.
6. (Cost) The total cost for the project shall not exceed $$.

Path	Event Sequence	S/C Contam-ination P.R. 1	S/C Rendez-vous P.R. 2	S/C Images P.R. 3	S/C Total Mass P.R. 4	S/C Launch Date P.R. 5	S/C Project Cost P.R. 6
1	M0 x N0 x A0 x B0 x C0 x D0 x E0				X	X	X
2	M1 x N0 x A0 x B0 x C0 x D0 x E0		X		X	X	X
3	M0 x N1 x A0 x B0 x C0 x D0 x E0				X	X	X
---	---	---	---	---	---	---	---
161	M0 x N0 x A0 x B0 x C0 x D0 x E2		X	X	X	X	X
162	M1 x N0 x A0 x B0 x C0 x D0 x E2		X	X	X	X	X
163	M0 x N1 x A0 x B0 x C0 x D0 x E2		X	X	X	X	X
---	---	---	---	---	---	---	---
254	M0 x N1 x A4 x B2 x C2	X	X	X	X	X	X
255	M1 x N1 x A4 x B2 x C2	X	X	X	X	X	X
256	M2		X	X	X	X	X

Continued

Inclusion of Mitigation Alternative 1 into the Expanded RSD for the Individual Risk Associated with Planet X Atmospheric Uncertainties from the Perspective of the Spacecraft Unit

4.4.2.2 Conducting the Risk Analysis and Documenting the Results

Conducting the risk analysis and documenting the results generally proceeds along the same lines as RIDM Step 4 of the same name. Each risk response alternative is represented by a separate risk analysis that quantifies the performance measures that would result if that alternative were to be implemented. From the performance measures and associated performance requirements, the performance risks of each alternative are calculated and serve as the primary results for risk-informing the selection of a risk response.

As in RIDM Step 4, the purpose of analyzing the risk response alternatives is to support decision making. The goal is a robust decision, where the decision-maker is confident that the selected risk response alternative is actually the best one, given the state of knowledge at the time. This requires the risk analysis to be rigorous enough to discriminate between alternatives, especially for those performance measures that are determinative to the decision. Therefore, it is expected that the analysis of risk response alternatives will typically be iterative, entailing additional analysis in areas where uncertainties hamper robust decision making.

The process of selecting a risk response alternative can be sequential, initially involving less detailed analysis in the level of rigor of assessed performance measures or only partial coverage of performance measures. Such initial analysis can help decision makers downselect contending risk response alternatives, which can then be analyzed further using more rigorous and/or comprehensive methods. Sensitivity studies can also be done to assure that the risk reduction potential of contending risk response alternatives are not unduly affected by modeling assumptions. This may be particularly useful for tactical risk response planning where point estimates are used in lieu of quantified uncertainties. In such cases it may be prudent to determine the range of parameter values over which the performance risk rankings of the most effective alternatives are stable, in order to give a qualitative sense of the robustness of the selection that the assessed risks support.

The communication tools of RIDM can be adapted for use in the presentation of risk analysis results here. Figure 63 and Figure 64 adapt Figure 30 and Figure 31 for application to the CRM analysis of risk response options.

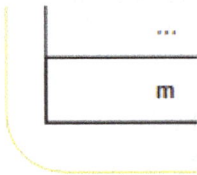

Figure 63. Notional Performance Risk Chart

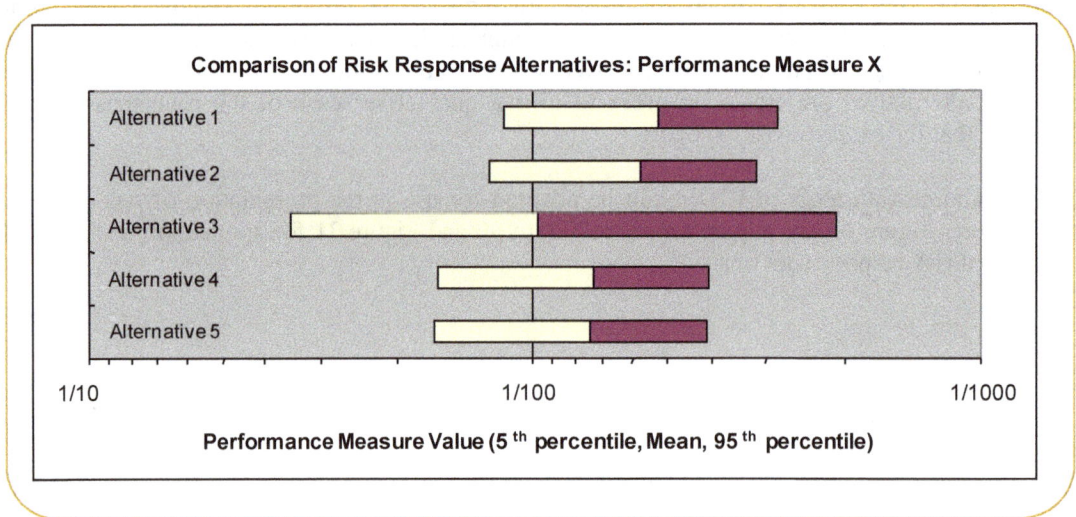

Figure 64. Notional Band-Aid Chart for Performance Measure X

4.4.3   Deliberation and Selection of a Risk Response

Deliberation and selection of a risk response generally proceeds along the same lines as RIDM Step 6. In parallel to that step, within CRM deliberation and selection of a risk response entails:

- **Convening a Deliberation Forum** – During implementation the deliberation forums that select risk responses should consist of personnel from both systems engineering and risk management, and should include stakeholder representation from every level of the NASA hierarchy that is substantially impacted not only by the performance risk that is being addressed by the risk response alternatives, but also by the performance risk that is produced by the risk response alternatives. For example, if one candidate means of addressing the performance risk to a technical requirement is to accept a higher performance risk to a cost or schedule requirement, those organizations that would be substantially impacted by the cost and schedule risk increase should be included in the deliberation forum.

- **Identify Contending Alternatives** – Risk response alternatives can be eliminated on the grounds of infeasibility, dominance, or inferior performance in key areas.

  - In the context of CRM, an infeasible alternative can be defined as one that fails to restore the performance risk to tolerable levels. Such an alternative fails to produce the desired risk reduction and can be eliminated in favor of alternatives that do. Infeasible alternatives should not be eliminated if there are no feasible alternatives. In other words, if the best alternatives fail to reduce the performance risk to tolerable levels, it still might be reasonable to select an alternative for implementation as a partial risk response, in tandem with elevation of further risk management decision making to the next higher level of the NASA hierarchy.

  - A dominated alternative can be defined as one whose performance risk impacts are inferior to those of another alternative for all performance requirements. Such an alternative can be eliminated in favor of the dominating alternative(s).

  - In general, some performance requirements may be of greater importance to the deliberators than others. Risk response alternatives that are markedly inferior to others in terms of their risk impact to key performance requirements can be eliminated on this basis, in recognition of decision-maker values.

- **Additional Uncertainty Considerations** – Performance risk addresses the probability that a given performance requirement will not be met, but it does not address the magnitude by which it may be exceeded (i.e., the full range of the uncertainty distribution for the performance measure). Such issues relate to the consideration of uncertainty beyond performance risk. In particular, a significant probability of exceptionally poor performance may be grounds for eliminating a risk response alternative from consideration. This may be the case even when the alternative returns all performance risks to tolerable levels if the consequences of failure are deemed sufficiently catastrophic. This issue is discussed in the framework of RIDM in Section 3.3.2.3.

- **Iteration with CRM *Analyze* and *Plan*** – As deliberation proceeds, a deliberator may have a particular issue or concern that requires additional analysis for closure. Consequently, deliberation is iterative with the CRM *Analyze* step, which can also produce a need for revised risk analyses of the risk response alternatives that are modeled in the CRM *Plan* step. Additionally, deliberators may suggest additional risk response alternatives. These might be modifications of existing alternatives, such as different combinations of existing risk response options, or alternatives that involve new options.

- **Communicating the Contending Alternatives to the Decision Maker** – The primary information that risk-informs the selection of a risk response alternative is the set of performance risks for each performance requirement, and the corresponding risk tolerances that the alternative achieves. Figure 65 illustrates a table of performance risks and tolerances where the time and cost to implement have been separated out. Additionally, information captured during deliberation should be summarized and forwarded to the decision-maker. This includes:

  o *The Pros and Cons of each Alternative* – An itemized table of the pros and cons of each alternative is recommended for the contending alternatives. It enables conflicting opinions to be documented, and captures elements of subjective value to the deliberators.

  o *Individual Risks introduced by each Alternative* – Risk response alternatives can potentially generate new individual risks as the cost of addressing existing performance risk. These should be identified and communicated to the decision-maker so that he or she understands the downside of each alternative.

- **Selecting a Risk Response Alternative** – Once the decision-maker has reviewed the risk information that has been provided to the deliberators, as well as the information generated during deliberation, he/she is in a position to make a risk-informed selection of a risk response alternative. The decision-maker should document the decision rationale, along with the information that supported the decision. A well-structured risk database should be able to generate a Risk Response Document (RRD) that contains the information relevant to a specific risk response decision.

Elevation of a Risk Decision – It may be the case that no risk response alternative is available that reduces performance risk to tolerable (or at least marginal) levels. In this case, elevation of the risk decision to the next level of the NASA organizational hierarchy is necessary. This situation would be documented, along with any other measures that are taken to at least partially address the intolerable performance risk.

- **Rebaselining of Performance Requirements** – There may also be situations that endanger the activity but are outside the CRM purview of accomplishing defined requirements. Examples of these include poorly defined or missing requirements and requirements creep. In such cases it may be necessary to return to the RIDM and System Engineering activities that lead to the derivation of performance requirements. The decision to rebaseline the requirements would be documented in the risk database and in the RRD that it generates.

Plan Alter-native No.	Time to Implement (and Uncertainty)	Cost to Implement (and Uncertainty)	Performance Risk Acceptability Results and Probability of Not Meeting Performance Requirement				Decision Maker's View of Whether Performance Risk Results are Acceptable, Marginal, or Unacceptable
			PM 1	PM 2	PM 3	---	
Null	0	0	$Pr = \ldots$	$Pr = \ldots$	$Pr = \ldots$		Unacceptable
1	$t_1 \pm \Delta t_1$	$C_1 \pm \Delta C_1$	$Pr = \ldots$	$Pr = \ldots$	$Pr = \ldots$		Acceptable because of requirement overlap
2	$t_2 \pm \Delta t_2$	$C_2 \pm \Delta C_2$	$Pr = \ldots$	$Pr = \ldots$	$Pr = \ldots$		Acceptable
3	$t_3 \pm \Delta t_3$	$C_3 \pm \Delta C_3$	$Pr = \ldots$	$Pr = \ldots$	$Pr = \ldots$		Marginal
4	$t_4 \pm \Delta t_4$	$C_4 \pm \Delta C_4$	$Pr = \ldots$	$Pr = \ldots$	$Pr = \ldots$		Unacceptable

Figure 65. Performance Risks and Risk Tolerances for the Contending Risk Response Alternatives

The following yellow box provides an example of how a decision between two mitigation alternatives could be informed by results of the risk analysis.

Planetary Science Mission Example: Analysis of Mitigation / Research Alternatives

The risk drivers for the planetary science mission example and the mitigation / research options recommended by the affected organizational units were identified in the yellow boxes within Section 4.4.1. As a result of deliberations within the project informed by these findings, the following two mitigation / research alternatives were selected for analysis:

1. Strengthen the four most windward thrusters in the RCS, incorporate an impact probe to obtain real-time atmospheric density measurements, replace some of the RTG power with ASRG power, and reallocate / redistribute the masses as needed for mass margin and center-of-gravity considerations.

2. Add thrusters to the RCS for redundancy, increase the heat shield thickness after performing tests to better characterize boundary layer transition and aerodynamic heating in high density environments, replace some of the RTG power with ASRG power, and reallocate/redistribute the masses as needed for mass margin and center-of-gravity considerations.

The RSDs, performance risk models, and associated input parameter distributions were updated to include each of these mitigation/research alternatives, and the graded approach risk analysis was performed to find the effect of each alternative on the performance risks. The results are shown in the table below (all numbers shown are hypothetical).

Plan Alter- native	Time to Implement (& Uncer- tainty)	Cost to Implement (& Uncertainty)	Performance Risk Tolerability & Probability of Not Meeting Performance Requirement						Decision Maker's View of Accep- tability
			No Planetary Contam- ination	Success- ful Rendez- vous	Video Images Meet Specs	S/C Mass within Require- ment	Launch Date within Window	Cost within Budget	
Null	0	0	P = 0.01	P = 0.02	P = 0.02	P = 0.005	P = 0.005	P = 0.05	TBD
1	8 mo. (± 3)	$0.5M (± 0.2)	P = 0.002	P = 0.004	P = 0.008	P = 0.006	P = 0.01	P = 0.008	TBD
2	6 mo. (± 2)	$1.0M (± 0.5)	P = 0.001	P = 0.002	P = 0.008	P = 0.02	P = 0.007	P = 0.008	TBD

Based on these results, Alternative 1 would be considered to be preferable to Alternative 2 since none of the performance risks are intolerable (although three are marginal). A negative is that the risk that the launch date will not be within the required window has increased from tolerable to marginal for Alternative 1 because of the need to design and test the impact probe along with its ejection mechanism and its control software.

Alternative 2 would be considered unacceptable at the present time because it is believed the masses required to augment the heat shield and the underlying spacecraft structure would be too large. However, positive research results from the boundary layer transition and aerodynamic heating tests in high density environments could reduce that risk. Thus the decision maker may opt to begin the conceptual designs for implementing Alternative 1 while conducting the research part of Alternative 2 to gather more information before making a final decision.

4.5   The CRM Track Step

The objective of the CRM *Track* step is twofold. It entails:

- Tracking the progress of the implementation of selected risk responses

- Tracking observables, related to performance measures and risk drivers, that are affected by the selected risk responses.

As such, the CRM *Track* step ensures that data are generated to monitor not only the implementation status of risk response options, but also their effectiveness once implemented. Figure 66 illustrates the process.

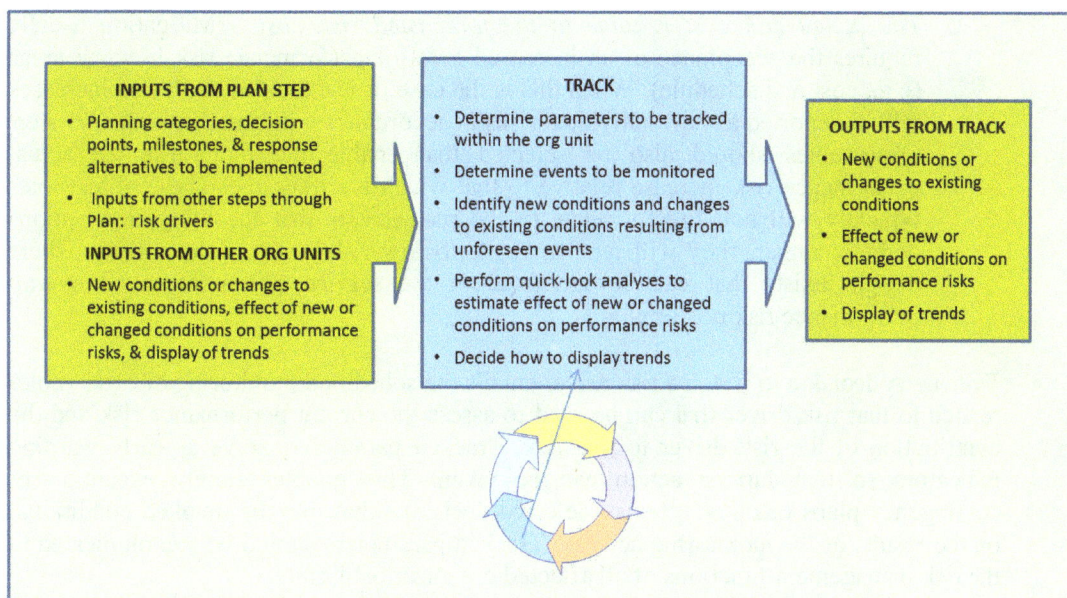

Figure 66. The CRM Track Step

Tracking applies to the *Mitigate*, *Watch*, and *Research* risk response option types. The option types *Accept*, *Close*, and *Elevate* do not have tracking requirements associated with them. The nature of tracking is a function of the option type for which the tracking is being performed:

- *Mitigate* – Mitigation produces a modification to the baseline project plan that reflects the implementation of the selected mitigation option(s). As such, implementation is expected to be integrated into the project schedule of the responsible organizational unit, and progress tracked by project management processes within that unit. The progress should be communicated to the risk management functions of other organizational units so that all affected risk management functions have an awareness of the current status/configuration of the activity.

Mitigation also entails the scheduled monitoring of observables related to the effectiveness of the mitigation option(s). These observable quantities should also be communicated to the risk management functions of other affected organizational units. Monitoring enables risk management to assess:

- O *The actual risk reduction relative to the forecasted risk reduction* – Mitigation options are implemented with the intent of reducing performance risk to the level forecasted by the risk analysis, conducted during the CRM *Plan* step, of the selected risk response alternative. Observables selected for tracking should enable the actual performance risk reduction to be assessed and compared to that forecasted during Plan. These observables are expected to be directly related to the risk drivers that the mitigation options address.

- O *The actual risk cost relative to the forecasted risk cost* – Mitigation usually requires the acceptance of an increased level of performance risk in some areas (e.g., cost and schedule). When this is the case, it is expected that these increases will be reported as new individual risks in accordance with the CRM *Identify* step. Observables should also be selected that enable the monitoring of actual performance risk increase relative to that which is forecasted. These observables typically will not directly relate to the risk drivers that the mitigation options address; rather, they will tend to relate to low-risk areas of the activity where margin exists that can be sacrificed in the service of an improved overall performance risk posture.

- *Watch* – A decision to watch a risk driver entails the scheduled monitoring of observables related to that risk driver that can be used to assess the current performance risk and the contribution of the risk driver to that risk. Tracked parameters serve as early warning indicators so that further action can be taken. This enables timely execution of contingency plans or other types of deferred decisions that may be invoked conditional on the results of the monitoring activity. Tracked parameters should be communicated to the risk management functions of all affected organizational units.

  In contrast to the *Mitigate* risk response option type, the *Watch* option type does not involve changes to the baseline project plan, and consequently does not involve the monitoring of implementation.

- *Research* – A decision to research a risk driver produces a research plan whose implementation should be tracked, and the scheduled monitoring of observables related to the research that, like *Watch*, can be used to assess the current performance risk and the contribution to that risk of the risk drivers associated with the research. Tracked parameters should be communicated to the risk management functions of all affected organizational units.

Tracking data can be used to construct performance risk tracking charts that show how performance risks increase and decrease over time as new individual risks are identified and responses are implemented. Figure 67 illustrates such a chart. The regions of risk tolerability are based on Figure 42 in Section 4.1.3.1.

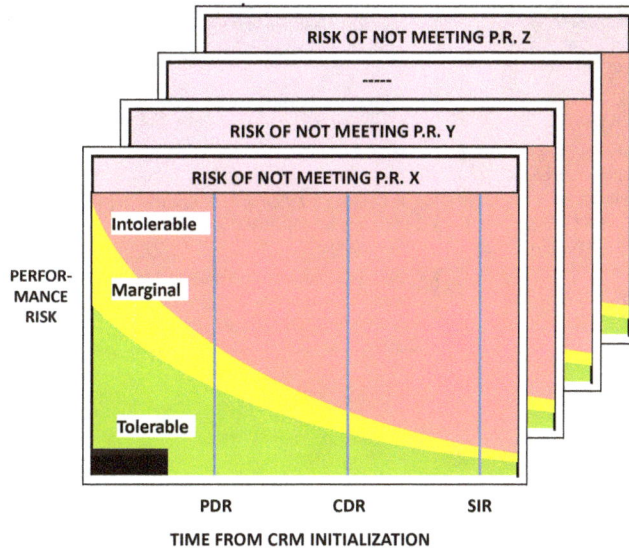

Figure 67. Performance Risk Tracking Chart

4.6   The CRM Control Step

The objective of the CRM *Control* step is to evaluate the tracking data to determine whether or not risk responses are being implemented as planned, and if so, whether or not they are effecting the anticipated changes in targeted risk drivers and in the performance risk generally. Control includes an assessment of the need to take action to keep the relevant risk responses on track. These actions are kept within the control function unless it is clear that the objective of the risk response cannot be attained within the current plan. If that is the case, the CRM *Plan* step is reinitiated and a new or modified risk response alternative is selected for implementation.

The CRM *Control* step is illustrated in Figure 68.

Because the *Control* step is focused on responding to the tracking data, it too is a function of risk response type:

- *Mitigate* – Because mitigation options are integrated into the project plan of the implementing organizational unit, implementation is overseen by the systems engineering function of the unit. Therefore, the role of risk management regarding control of implementation is primarily one of monitoring progress and evaluating the potential risk consequences associated with departures from the implementation plan. It also includes implementing contingencies when needed and making small changes that do not require reinitiation of the Plan step.

185 of 234

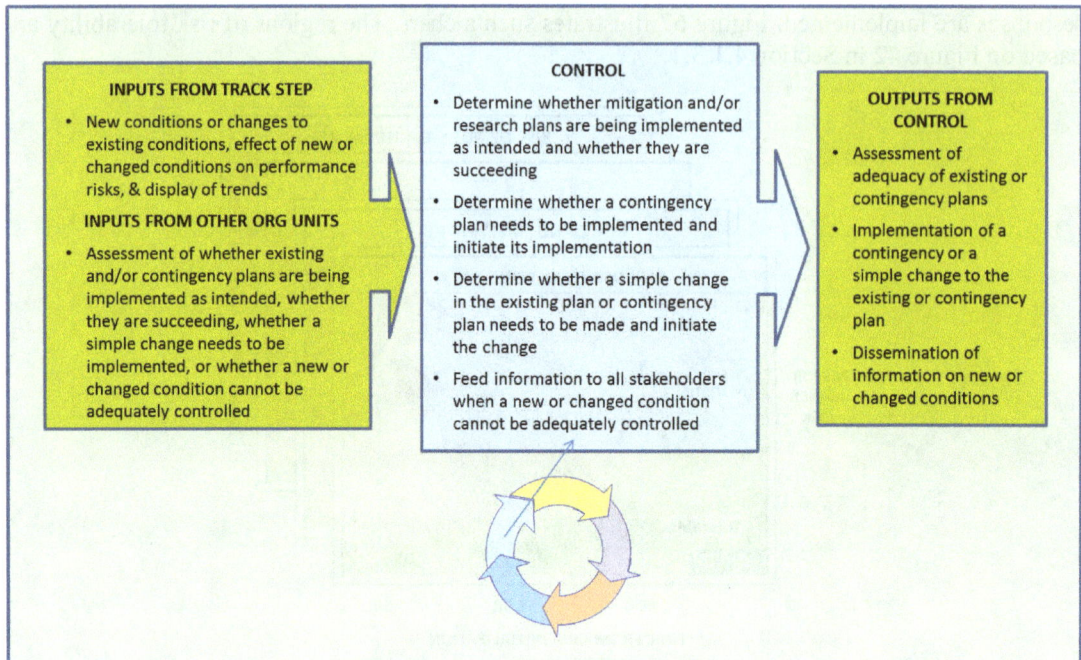

Figure 68. The CRM Control Step

As mitigation options are implemented, it is the function of the CRM *Control* step to evaluate the updated risk model in light of the tracked parameters and assess the degree to which they have successfully mitigated the effects of the risk driver(s) they address. If the assessed performance risk falls short of that forecasted during *Plan*, *Control* acts within the scope of the selected alternative to achieve, at least approximately, the intended result.

- *Watch* – In the case of *Watch* options, the CRM *Control* step evaluates the watched parameters and, as appropriate, executes the contingency plans or other deferred decisions according to pre-established criteria.

- *Research* – Like *Mitigate*, *Research* involves the execution of a plan of action (in this case, the research plan) whose implementation is expected to be integrated into the project plan. Therefore, the role of risk management regarding the control of research option implementation is analogous to that for mitigation.

  Like *Watch*, *Research* involves the execution of contingency plans or other deferred decisions based on an evaluation of the researched parameters relative to pre-established criteria.

4.7 Communicate and Document

As illustrated in the graphic of Figure 39, communication and documentation are central to CRM, and are integrated into each of the five CRM steps of *Identify*, *Analyze*, *Plan*, *Track*, and *Control*. Each of these steps in the CRM process involves the generation of information that must be properly documented and communicated to the appropriate personnel at the appropriate time, using appropriate standardized communication aids to assure that the intended meaning has been conveyed.

4.7.1 Communication Within CRM

Throughout the CRM process, communication takes place among stakeholders involved in risk management, project management, and systems engineering to make sure that risks are effectively managed during implementation. As discussed in previous subsections, communication can take place in a variety of forums, ranging from informal meetings, phone calls, and emails among personnel within an organizational unit, to technical interchange meetings involving personnel from numerous units in the NASA hierarchy and the authoring and dissemination of detailed reports. A graded approach is appropriate to determining the scale and formality of a given forum; in general, forums should be used that facilitate dissemination of information to the relevant affected parties, and provide ample opportunity for discussion and feedback to assure that issues are fully understood at a level that supports the decision making needs of all participants.

Inter-organizational communication is an integral part of CRM in the context of the NASA organizational hierarchy. Different organizational units at different levels in the hierarchy must work together to ultimately achieve the top-level objectives that motivate their derived requirements. Throughout the CRM process, communication takes place among these units to assure that:

- Every unit is aware of the individual risks that affect its performance risk.

- Individual risks are integrated into the risk analyses of the affected units in a consistent fashion (i.e., using consistent modeling assumptions).

- Every unit's risk driver lists is available to other units and is updated according to an established schedule.

- Every unit that is affected by a risk driver, or by the proposed responses to a risk driver, are adequately engaged in planning a response to it, including deliberation and selection of a response for implementation.

- Every unit is aware of the risk responses that affect its performance risk and/or its risk analysis.

- Elevation of risk management decisions is timely and unambiguous.

Standardized communication aids should be developed that support the information needs of the decisions they support. Examples are provided in many of the previous CRM sections of graphical aids such as:

- Risk burn-down schedules for each performance requirement, indicating the tolerability of performance risk as a function of project phase for each performance requirement

- Populated risk taxonomies showing the distribution of individual risks among the taxons of the three specified taxonomies (condition/departure, asset, and consequence)

- RSDs that enumerate the spectrum of possible outcomes (and their likelihoods) resulting from an individual risk's departure event

- Criticality rank tables that support tactical and strategic risk analysis

- Risk driver lists to support risk response planning

- Tables of performance risk and risk tolerances for the contending risk response alternatives

- Risk tracking aids such as performance risk tracking charts that show the trajectories of each performance risk with respect to its risk burn-down profile.

Risk communication protocols should be negotiated among involved organizational units and documented in the RMP. This includes scheduled periodic reporting of risk information, such as to the unit at the next higher level, as well as protocols for risk reporting in response to triggers such as the exceedance of an elevation threshold.

4.7.2 Documentation within CRM

The risk database can be used as the central repository of risk management documentation related to CRM. In order to fully support the CRM process, the risk database must be relational, allowing for many-to-many linkages between individual risks, performance risks, risk drivers, and risk responses. It provides storage and archiving of the risk analysis results as they evolve over the course of the activity. It also provides storage and archiving of risk response planning, including the set of risk response alternatives, the risk analyses of the alternatives, the selected risk response, and the rationale for the selection.

4.8 Applicability of Project-Centered CRM Processes to Other Risk Domains

The intent of this section is to discuss risks that have not been considered in the handbook to this point but that impact (1) the ability of the Agency to meet its goals and (2) the ability of the programs/projects within the Agency to meet their requirements. Some thoughts are presented as to how the methods and processes discussed within the handbook might apply to those risks, even though it is recognized that additional investigation is needed to recommend methods for treating them in a holistic fashion together with the risks that have already been discussed.

4.8.1   Institutional Risks

As mentioned in the Section 4 introductory remarks, institutional risks are defined in NPR 8000.4A as follows: "Risks to infrastructure, information technology, resources, personnel, assets, processes, occupational safety, environmental management, or security that affect capabilities and resources necessary for mission success, including institutional flexibility to respond to changing mission needs and compliance with external requirements (e.g., Environmental Protection Agency or Occupational Safety and Health Administration regulations)." They include risks that affect the strength and mix of the workforce, the viability and maintenance of facilities, the ability to meet QA standards, the effectiveness and accessibility of support functions, the usefulness and maintainability of software that supports day-to-day operations, and other related issues. In accordance with the NPR 8000.4A definition, institutional risks are separated from enterprise risks and Agency-wide strategic risks, which are discussed separately in Sections 4.8.2 and 4.8.3.

Institutional risks can be handled in much the same way as project risks, as long as there is a responsible organizational unit for each risk. The responsible organization has to have its own set of performance requirements and goals and its own set of performance metrics by which it can measure its success or shortfalls. As long as the performance metrics can be expressed quantitatively, they can be treated using the same methods and processes discussed in this handbook.

The issue that requires further analysis is how to model the effects of institutional risks on project risks. For example, if there is something that could cause the maintenance schedule for software to lag behind the planned schedule, how does this affect the probability that the software might fail or become obsolete, how long might it take for this to happen, and how would this affect the ability of the project to meet its milestones for those organizational units that depend upon the software? In principle, the means for analyzing these dependencies should be similar to what has already been discussed. The specifics, however, may require special considerations as to the relationships between the software and the project functions, the interfaces between the institutional organization and the project organizations, and the competition for support services among different projects within the Agency.

4.8.2   Enterprise Risks

Enterprise risk management at NASA involves the management of risks that pertain to the operation of the Agency as a business or corporation. Like project risk management, the focus is on identifying particular events or circumstances relevant to the organization's objectives (risks and opportunities), systematically analyzing them in terms of likelihood and magnitude of impact, planning a response strategy, tracking progress, and controlling deviations.

Like project risks, enterprise risks are often grouped into types or domains. Not surprisingly, the domains are defined differently for the two categories of risk because the nature of the risks differs for each. Below is a list of five domains that are sometimes used to categorize enterprise risks: [38]

- Acquisition risks, including risks that cause a hindering of cohesive teamwork or tend to concentrate market power in a single entity

- Strategic risks, including regulatory and political issues, obsolescence due to technical advances, reputational damage, changes in stakeholder wants and preferences, and lack of capital availability

- Financial risks, including commodity risks (e.g., changes in price of raw materials), credit risk (e.g., defaults on accounts receivable), and inadequate liquidity

- Operational risks, including availability of human resources, leadership transitions, efficiency and reliability of machines and processes, and supply chain management

- Hazard risks, including external events (e.g., fire, theft, acts of terror), injuries and disabilities, and liability suits.

Many of the processes for managing project risks carry over to the management of enterprise risks. In addition, there are many textbooks and software tools aimed at enterprise risk management for businesses. The application of these processes, texts, and tools for an agency like NASA, however, is an area that needs to be further investigated.

4.8.3 Agency-Wide Strategic Risks

At the Agency level, there is a set of strategic risk issues that are ranked against narrative performance goals. For example, one of the strategic goals (No. 3A) in the NASA FY 2010 Performance Plan is to "study Earth from space to advance scientific understanding and meet societal needs." The performance over the past fiscal year for that goal is currently ranked according to outcomes that contribute to the performance goals. For example, one of the outcomes (No. 3A.5) is defined as "progress in understanding the role of oceans, atmosphere, and ice in the climate system and in improving predictive capability for its future evolution." Each outcome is judged on the basis of annual performance goals (APGs). For example, one of the APGs (No. 10ES12) for this particular outcome is to "develop missions as demonstrated by completing the ICESat-II Initial Confirmation Review." Some outcomes and APGs are specifically related to cross-Agency support functions and/or overall uniformity and efficiency goals. For example, outcome ED.1 is defined as "making a contribution to the development of the Science, Technology, Engineering, and Math (STEM) workforce in disciplines that are needed to achieve NASA's Strategic Goals, through a portfolio of investments." Within that outcome, APG 10ED01 is defined as "supporting the development of 60 new or revised courses targeted at the STEM skills."

Performance against outcomes in the Performance Plan is ranked Green, Yellow, or Red, according to the following criteria:

- Green: NASA achieved most APGs under this outcome and is on-track to achieve or exceed this outcome.

- Yellow: NASA made significant progress toward this outcome, however, the Agency may not achieve this outcome as stated.

- Red: NASA failed to achieve most of the APGs under this outcome and does not expect to achieve this outcome as stated.

While the present process is focused on past performance, a similar process can be used to manage risk with respect to future performance. The focus in that case would be on considering the outcomes and APGs that are expected to be applicable in the future. The process for managing risk with a future orientation could be quite similar to the process for managing project risks. For example, one could define the severity of the strategic risk as having three discrete levels, as follows:

- Least severe: The risk issue may cause NASA to miss a small percentage of applicable APGs but NASA will stay on track for achieving all the applicable outcomes under the goal.

- Moderately severe: The risk issue may cause NASA to miss a moderate percentage of applicable APGs such that NASA may not be able to achieve all its applicable outcomes as stated, but progress toward achieving all the applicable outcomes will be significant.

- Most severe: The risk issue may cause NASA to fail to achieve most of the applicable APGs and to not achieve one or more of the applicable outcomes under the goal.

At the outset of the risk management process, the appropriate decision maker would specify his or her tolerability for likelihoods of each severity level as a function of time, using a color-coded format such as that in Figure 42. Each strategic goal could have different tolerability specifications, similar to how each performance requirement has different tolerability specifications in Figure 42.

The details for addressing Agency strategic risk remain to be further developed, but it would seem plausible that the risk management treatment for the Agency's strategic goals could follow a path similar to the risk management treatment for project risks.

5   REFERENCES

1.   NASA. *NPR 8000.4A, Agency Risk Management Procedural Requirements.* Washington, DC. 2008.

2.   NASA. *NASA/SP-2007-6105, NASA Systems Engineering Handbook.* Washington, DC. 2007.

3.   NASA. *NPR 7123.1A, NASA Systems Engineering Processes and Requirements.* Washington, DC. 2007.

4.   NASA. *NPR 7120.5D, NASA Space Flight Program and Project Management Processes and Requirements.* Washington, DC. 2007.

5.   NASA. *NPR 7120.7, NASA Information Technology and Institutional Infrastructure Program and Project Management Requirements,* Washington, DC. 2008.

6.   NASA. *NPR 7120.8, NASA Research and Technology Program and Project Management Requirements.* Washington, DC. 2008.

7.   NASA. *NPD 1000.5, Policy for NASA Acquisition.* Washington, DC. 2009.

8.   NASA. *NPD 7120.4C, Program/Project Management. Washington,* DC. 1999.

9.   NASA. *NPD 8700.1, NASA Policy for Safety and Mission Success.* Washington, DC. 2008.

10.   Dezfuli, H., Stamatelatos, M., Maggio, G., and Everett, C., "Risk-informed Decision Making in the Context of NASA Risk Management," PSAM 10, Seattle, WA, 2010.

11.   NASA. *NPD 1000.0A, Governance and Strategic Management Handbook.* Washington, DC. 2008.

12.   Hammond, J., Keeney, R., and Raiffa, H. "The Hidden Traps in Decision Making." *Harvard Business Review,* September – October 1998.

13.   Carnegie Mellon University Software Engineering Institute. *Continuous Risk Management Guidebook.* 1996.

14.   Clemen, R., *Making Hard Decisions.* Pacific Grove, CA. Duxbury Press, 1996.

15.   Keeney, R., and Raiffa, H., *Decisions with Multiple Objectives: Preferences and Value Tradeoffs.* Cambridge, UK: Cambridge University Press, 1993.

16.   Hammond, J., Keeney, R., and Raiffa, H., "Even Swaps: A Rational Method for Making Trade-offs." *Harvard Business Review.* March – April 1998.

17.   NRC. *Understanding Risk – Informing Decisions in a Democratic Society.* The National Academies Press, 1996.

18.   U.S. Forest Service, Pacific Southwest Research Station. *Comparative Risk Assessment Framework and Tools (CRAFT),* Version 1.0, 2005. (http://www.fs.fed.us/psw/topics/fire_science/craft/craft/index.htm)

19.   NASA Aerospace Safety Advisory Panel. "Aerospace Safety Advisory Panel Annual Report for 2009," Washington, D.C., 2010.

20.   NASA. "Constellation Program Implementation of Human-Rating Requirements," Tracking Number 2009-01-02a, Office of the Administrator, Washington, D.C., 2010.

21.   Keeney, R., *Value-Focused Thinking: A Path to Creative Decisionmaking*, Harvard University Press, 1992.

22.   Keeney, R., and McDaniels, T., "A Framework to Guide Thinking and Analysis Regarding Climate Change Policies," *Risk Analysis* 21, No. 6, pp. 989-1000, Society for Risk Analysis, 2001.

23.   NASA. "Exploration Systems Architecture Study -- Final Report," NASA-TM-2005-214062, Washington, DC. 2005.

24.   Maggio, G., Torres, A., Keisner, A., and Bowman, T., "Streamlined Process for Assessment of Conceptual Exploration Architectures for Informed Design (SPACE-AID)," Orlando, FL. 2005.

25.   NASA. *NPR 8715.3C, NASA General Safety Program Requirements.* Washington, DC. 2008.

26.   NASA. *NASA Cost Estimating Handbook.* Washington, DC. 2008.

27.   NASA. *Probabilistic Risk Assessment Procedures Guide for NASA Managers and Practitioners*, Version 1.1. Washington, DC. 2002.

28.   Morgan, M., and Henrion, M., *Uncertainty: A Guide to Dealing with Uncertainty in Quantitative Risk and Policy Analysis*, Cambridge Press, 1990.

29.   Apostolakis, G., "The Distinction between Aleatory and Epistemic Uncertainties is Important: An Example from the Inclusion of Aging Effects into Probabilistic Safety Assessment," Washington, DC. 1999.

30.   NASA. *NASA/SP-2009-569, Bayesian Inference for NASA Probabilistic Risk and Reliability Analysis.* Washington, DC. 2009.

31. Mosleh, A., Siu, N., Smidts, C., and Lui, C., "Model Uncertainty: Its Characterization and Quantification," University of Maryland, 1993.

32. Attoh-Okine, N., and Ayyub, B., "Applied Research in Uncertainty Modeling and Analysis," International Series in Intelligent Technologies, Springer, 2005.

33. NASA. *NASA-STD-7009, Standard for Models and Simulations*. Washington, DC. 2008.

34. Groen, F., and Vesely, B., "Treatment of Uncertainties in the Comparison of Design Option Safety Attributes," PSAM 10, Seattle, WA, 2010.

35. Stromgren, C., Cates, G., and Cirillo, W., "Launch Order, Launch Separation, and Loiter in the Constellation 1½-Launch Solution," SAIC/NASA LRC, 2009.

36. Bearden, D., Hart, M., Bitten, R., et al, "Hubble Space Telescope (HST) Servicing Analyses of Alternatives (AoA) Final Report," The Aerospace Corporation, 2004.

37. Guarro, S., "Goals and Reality of Risk Management: Lessons Learned from Space Program Executions," Proceedings of Space Systems Engineering & Risk Management Symposium, Oct. 26-28, 2005.

38. Casualty Actuarial Society, "Overview of Enterprise Risk Management," Report of the Enterprise Risk Management Committee, May 2003.

APPENDIX A: ACRONYMS AND ABBREVIATIONS

AHP	Analytic Hierarchy Process
AoA	Analysis of Alternatives
ASAP	Aerospace Safety Advisory Panel
CAS	Credibility Assessment Scale
CCDF	Complementary Cumulative Distribution Function
CD	Center Director
CDR	Critical Design Review
CEV	Crew Exploration Vehicle
CFD	Computational Fluid Dynamics
Ci	Curie
CLV	Crew Launch Vehicle
ConOps	Concept of Operations
COS	Cosmic Origins Spectrograph
CRAFT	Comparative Risk Assessment Framework and Tools
CRM	Continuous Risk Management
DB	Database
DDT&E	Design, Development, Test & Evaluation
DM	De-orbit Module
DMS	Document Management System
DoD	Department of Defense
DRM	Design Reference Mission
EDS	Earth Departure Stage
EELV	Evolved Expendable Launch Vehicle
EOL	End of Life
EOM	End of Mission
ESAS	Exploration Systems Architecture Study
ETO	Earth-to-Orbit
FGS	Fine Guidance Sensor
FOM	Figure of Merit
G&A	General and Administrative
HQ	Headquarters
HST	Hubble Space Telescope
ISS	International Space Station
JCL	Joint Confidence Level
KSC	Kennedy Space Center

LCF	Latent Cancer Fatality
LEO	Low Earth Orbit
LLO	Low Lunar Orbit
LOC	Loss of Crew
LOI	Lunar Orbit Insertion
LOM	Loss of Mission
LSAM	Lunar Surface Access Module
M&S	Modeling & Simulation
MAUT	Multi-Attribute Utility Theory
MMRTG	Multi-Mission Radioisotope Thermoelectric Generator
MOE	Measure of Effectiveness
MSFC	Marshall Space Flight Center
MSO	Mission Support Office
NASA	National Aeronautics and Space Administration
NPD	NASA Policy Directive
NPR	NASA Procedural Requirements
NRC	Nuclear Regulatory Commission
NRC	National Research Council
ODC	Other Direct Cost
PC	Performance Commitment
pdf	Probability Density Function
PDR	Preliminary Design Review
PM	Performance Measure
pmf	Probability Mass Function
PnSL	Probability of No Second Launch
PR	Performance Requirement
PRA	Probabilistic Risk Assessment
Pu	Plutonium
RCS	Reaction Control System
RIDM	Risk-Informed Decision Making
RISR	Risk-Informed Selection Report
RM	Risk Management
RMP	Risk Management Plan
ROM	Rough Order-of-Magnitude
RRP	Risk Response Plan
RRW	Risk Reduction Worth
RSD	Risk Scenario Diagram
RSRB	Redesigned Solid Rocket Booster
RTG	Radioisotope Thermoelectric Generator

S&MA	Safety & Mission Assurance
SDM	Service and De-orbit Module
SDR	System Design Review
SEI	Software Engineering Institute at Carnegie Mellon University
SIR	System Integration Review
SM	Service Module
SME	Subject Matter Expert
SMn	Servicing Mission n
SRR	System Requirements Review
SSME	Space Shuttle Main Engine
STS	Space Transportation System
SW	Software
TBfD	Technical Basis for Deliberation
TLI	Trans-Lunar Injection
TRL	Technology Readiness Level
WBS	Work Breakdown Structure
WFC3	Wide Field Camera 3

APPENDIX B: DEFINITIONS

<u>Aleatory</u>: Pertaining to stochastic (non-deterministic) events, the outcome of which is described by a pdf. From the Latin alea (game of chance, die). [Adapted from [30]]

<u>Consequence</u>: The possible negative outcomes of the current conditions that are creating uncertainty. [Adapted from [13]]

<u>Continuous Risk Management (CRM)</u>: A specific process for the management of risks associated with implementation of designs, plans, and processes. The CRM functions of identify, analyze, plan, track, control, and communicate and document provide a disciplined environment for continuously assessing what could go wrong, determining which issues are important to deal with, and implementing strategies for dealing with them. [Adapted from [13]]

<u>Deliberation</u>: Any process for communication and for raising and collectively considering issues. In deliberation, people discuss, ponder, exchange observations and views, reflect upon information and judgments concerning matters of mutual interest, and attempt to persuade each other. Deliberations about risk often include discussions of the role, subjects, methods, and results of analysis. [Excerpted from [17]]

<u>Dominated Alternative</u>: An alternative that is inferior to some other alternative with respect to every performance measure.

<u>Epistemic</u>: Pertaining to the degree of knowledge. From the Greek episteme (knowledge). [Adapted from [30]]

<u>Imposed Constraint</u>: A limit on the allowable values of the performance measure with which it is associated. Imposed constraints reflect performance requirements that are negotiated between NASA organizational units and which define the task to be performed.

<u>Likelihood</u>: Probability of occurrence.

<u>Objective</u>: A specific thing that you want to achieve. [14]

<u>Performance Commitment</u>: A level of performance that a decision alternative is intended to achieve at a given level of risk. Performance commitments are established on the performance measures of each alternative as a means of comparing performance across alternatives that is consistent with the alternative-independent risk tolerance of the decision-maker.

Performance Measure: A metric used to measure the extent to which a system, process, or activity fulfills its associated performance objective. [Adapted from [1]]

Performance Objective: An objective whose fulfillment is directly quantified by an associated performance measure. In the RIDM process, performance objectives are derived via an objectives hierarchy, and represent the objectives at the levels of the hierarchy.

Performance Parameter: Any value needed to execute the models that quantify the performance measures. Unlike performance measures, which are the same for all alternatives, performance parameters typically vary among alternatives, i.e., a performance parameter that is defined for one alternative might not apply to another alternative.

Performance Requirement: The value of a performance measure to be achieved by an organizational unit's work that has been agreed-upon to satisfy the needs of the next higher organizational level. [1]

Quantifiable: An objective is quantifiable if the degree to which it is satisfied can be represented numerically. Quantification may be the result of direct measurement; it may be the product of standardized analysis; or it may be assigned subjectively.

Risk: In the context of RIDM, risk is the potential for shortfalls, which may be realized in the future, with respect to achieving explicitly-stated performance commitments. The performance shortfalls may be related to institutional support for mission execution, or related to any one or more of the following mission execution domains: safety, technical, cost, schedule.

As applied to CRM, risk is characterized as a set of triplets:

    a. The scenario(s) leading to degraded performance in one or more performance measures,

    b. The likelihood(s) of those scenarios,

    c. The consequence(s), impact, or severity of the impact on performance that would result if those scenarios were to occur.

Uncertainties are included in the evaluation of likelihoods and consequences. [Adapted from [1]]

Risk Analysis: For the purpose of this handbook, risk analysis is defined as the probabilistic assessment of performance such that the probability of not meeting a particular performance commitment can be quantified.

Risk Averse: The risk attitude of preferring a definite outcome to an uncertain one having the same expected value.

Risk-Informed Decision Making: A risk-informed decision-making process uses a diverse set of performance measures (some of which are model-based risk metrics) along with other considerations within a deliberative process to inform decision making.

> *Note: A decision-making process relying primarily on a narrow set of model-based risk metrics would be considered "risk-based."* [1]

Risk Management: Risk management includes RIDM and CRM in an integrated framework. This is done in order to foster proactive risk management, to better inform decision making through better use of risk information, and then to more effectively manage implementation risks by focusing the CRM process on the baseline performance requirements emerging from the RIDM process. [1]

Risk Seeking: The risk attitude of preferring an uncertain outcome to a certain one having the same expected value.

Robust: A robust decision is one that is based on sufficient technical evidence and characterization of uncertainties to determine that the selected alternative best reflects decision-maker preferences and values given the state of knowledge at the time of the decision, and is considered insensitive to credible modeling perturbations and realistically foreseeable new information.

Scenario: A sequence of credible events that specifies the evolution of a system or process from a given state to a future state. In the context of risk management, scenarios are used to identify the ways in which a system or process in its current state can evolve to an undesirable state.

Sensitivity Study: The study of how the variation in the output of a model can be apportioned to different sources of variation in the model input and parameters. [33]

Stakeholder: A stakeholder is an individual or organization that is materially affected by the outcome of a decision or deliverable.

Uncertainty: An imperfect state of knowledge or a physical variability resulting from a variety of factors including, but not limited to, lack of knowledge, applicability of information, physical variation, randomness or stochastic behavior, indeterminacy, judgment, and approximation. [1]

APPENDIX C: CONTENT GUIDE FOR THE TECHNICAL BASIS FOR DELIBERATION

Technical Basis for Deliberation Content

The Technical Basis for Deliberation (TBfD) document is the foundation document for the risk-informing activities conducted during Part 1 and Part 2 of the RIDM Process. The TBfD conveys information on the performance measures and associated imposed constraints for the analyzed decision alternatives.

Because the TBfD provides the specific risk information to understand the uncertainty associated with each alternative, this document serves as the technical basis for risk-informed selection of alternatives within the program or project. The risk analysis team, working under the overall program/project guidance, develops TBfD documentation and updates the information provided as necessary based upon questions and/or concerns of stakeholders during deliberation. The risk analysis team works with the deliberators and decision-maker to support deliberation and alternative selection.

The TBfD includes the following general sections:

- Technical Summary: This section describes the problem to be solved by this effort and each of the general contexts of each of the alternatives.

- Top-level Requirements and Expectations: This section contains the top-level requirements and expectations identified in Step 1 of the RIDM process. In cases involving diverse stakeholders, a cross reference between expectations and stakeholder may be presented.

- Derivation of Performance Measures: This section shows the derivation of performance measures for the decision conducted in Step 1 of the RIDM process. Typical products are the objectives hierarchy and a table mapping the performance objectives to the performance measures. When proxy performance measures are used, their definitions are provided along with the rationale for their appropriateness. When constructed scales are used, the scales are presented.

- Decision Alternatives: This section shows the compilation of feasible decision alternatives conducted in Step 2 of the RIDM process. Typical products are trade trees, including discussion of tree scope and rationales for the pruning of alternatives prior to risk analysis. Alternatives that are retained for risk analysis are described. This section also identifies any imposed constraints on the allowable performance measure values, and a map to the originating top-level requirements and/or expectations.

- Risk Analysis Framework and Methods: This section presents the overall risk analysis framework and methods that are set in Step 3 of the RIDM process. For each analyzed alternative, it shows how discipline-specific models are integrated into an analysis process that preserves correlations among performance parameters. Discipline-specific analysis models are identified and rationale for their selection is given. Performance parameters are identified for each alternative.

- Risk Analysis Results: This section presents the risk analysis results that are quantified in Step 4 of the RIDM process.

    o Scenario descriptions: For each alternative, the main scenarios identified by the risk analysis are presented.

    o Performance measure pdfs: For each alternative, the marginal performance measure pdfs are presented, along with a discussion of any significant correlation between pdfs.

    o Imposed constraint risk: For each alternative, the risk with respect to imposed constraints is presented, along with a discussion of the significant drivers contributing to that risk.

    o Supporting analyses: For each alternative, uncertainty analyses and sensitivity studies are summarized.

- Risk Analysis Credibility Assessment: This sections presents the credibility assessment performed in accordance with [33].

APPENDIX D: CONTENT GUIDE FOR THE RISK-INFORMED SELECTION REPORT

Risk-Informed Selection Report Content

The Risk-Informed Selection Report (RISR) documents the rationale for selection of the selected alternative and demonstrates that the selection is risk-informed. The decision-maker, working with the deliberators and risk analysis team, develops the RISR.

The RISR includes the following general sections:

- Executive Summary: This summary describes the problem to be solved by this effort and each of the general contexts of each of the alternatives. It identifies the organizations and individuals involved in the decision-making process and summarizes the process itself, including any intermediate downselects. It presents the selected alternative and summarizes the basis for its selection.

- Technical Basis for Deliberation: This section contains material from the TBfD (see Appendix C).

- Performance Commitments: This section presents the performance measure ordering and risk tolerances used to develop the performance commitments during Step 5 of the RIDM process, with accompanying rationale. It tabulates the resultant performance commitments for each alternative.

- Deliberation: This section documents the issues that were deliberated during Step 6 of the RIDM process.

    o Organization of the deliberations: The deliberation and decision-making structure is summarized, including any downselect decisions and proxy decision-makers.

    o Identification of the contending decision alternatives: The contending alternatives are identified, and rationales given for their downselection relative to the pruned alternatives. Dissenting opinions are also included.

    o Pros and cons of each contending alternative: For each contending alternative, its pros and cons are presented, along with relevant deliberation issues including dissenting opinions. This includes identifying violations of significant engineering standards, and the extent to which their intents are met by other means.

    o Deliberation summary material: Briefing material, etc., from the deliberators and/or risk analysts to the decision-maker (or decision-makers, in the case of multiple downselects) is presented.

- <u>Alternative Selection</u>: This section documents the selection of an alternative conducted in Step 6 of the RIDM process.

  - <u>Selected alternative</u>: The selected alternative is identified, along with a summary of the rationale for its selection.

  - <u>Performance commitments</u>: The finalized performance commitments for the selected alternative are presented, along with the final performance measure risk tolerances and performance measure ordering used to derive them.

  - <u>Risk list</u>: The RIDM risk list for the selected alternative is presented, indicating the risk-significant conditions extant at the time of the analysis, and the assessed impact on the ability to meet the performance commitments.

  - <u>Decision robustness</u>: An assessment of the robustness of the decision is presented.

APPENDIX E: SELECTED NASA EXAMPLES OF RIDM PROCESS ELEMENTS

NASA has a long history of incorporating risk considerations into its decision-making processes. As part of the development of this handbook, NASA OSMA reviewed a number of decision forums and analyses for insights into the needs that the handbook should address, and for examples of decision-making techniques that are illustrative of elements of the resultant RIDM process.

The following example process elements are intended as illustrations of the general intent of the RIDM process elements to which they correspond. They do not necessarily adhere in every detail to the guidance in this handbook. Nevertheless, they represent sound techniques that have risk-informed decision making at NASA.

E.1   Stakeholder Expectations

E.1.1   The Use of Design Reference Missions in ESAS [23]

A series of DRMs was established to facilitate the derivation of requirements and the allocation of functionality between the major architecture elements. Three of the DRMs were for missions to the International Space Station (ISS): transportation of crew to and from the ISS, transportation of pressurized cargo to and from the ISS, and transportation of unpressurized cargo to the ISS. Three of the DRMs were for lunar missions: transportation of crew and cargo to and from anywhere on the lunar surface in support of 7-day "sortie" missions, transportation of crew and cargo to and from an outpost at the lunar south pole, and one-way transportation of cargo to anywhere on the lunar surface. A DRM was also established for transporting crew and cargo to and from the surface of Mars for an 18-month stay. Figures E-1 and E-2 show two of the ESAS DRMs: one for an ISS mission and one for a lunar mission.

Figure E-1. Crew Transport to and from ISS DRM

Figure E-2. Lunar Sortie Crew with Cargo DRM

E.2   Objectives Hierarchies and Performance Measures

E.2.1   The Use of Figures of Merit (FOMs) in ESAS [23]

The various trade studies conducted by the ESAS team used a common set of FOMs (a.k.a. performance measures) for evaluation.[32] Each option was quantitatively or qualitatively assessed against the FOMs shown in Figure E-3. FOMs were included in the areas of: safety and mission success, effectiveness and performance, extensibility and flexibility, programmatic risk, and affordability. FOMs were selected to be as mutually exclusive and measurable as possible.

---

[32] The inclusion of this example should not be taken as advocating any particular set of performance measures. In particular, the treatment of risk in terms of explicit risk FOMs is inconsistent with the RIDM process as discussed in Section 3.1.2.

Figure E-3. ESAS FOMs

E.2.2   The Use of Figures of Merit in "Launch Order, Launch Separation, and Loiter in the Constellation 1½-Launch Solution" [35]

The goal of this launch order analysis was to evaluate the identified operational concepts and then produce a series of relevant FOMs for each one. The most basic metric that was considered was the probability that each concept would result in a failure to launch the second vehicle. The FOMs for the study had to cover a number of areas that were significant to decision-makers on selecting a concept. Table E-1 lists the FOMs that were considered in this study.

Table E-1. Launch Order Risk Analysis FOMs

FOMs
Probability of No Second Launch
Cost of Failure
Loss of Delivery Capability to the Surface
Additional Risk to the Crew
Additional Costs or Complexities

E.3　Compiling Alternatives

E.3.1　The Use of Trade Trees in ESAS [23]

Figure E-4 shows the broad trade space for Earth-to-orbit (ETO) transportation defined during ESAS. In order to arrive at a set of manageable trade options, external influences, as well as technical influences, were qualitatively considered in order to identify feasible decision alternatives.

Figure E-4. Possible Range of ESAS Launch Trade Study

The decision points, illustrated numerically in Figure E-4, are described below, with the subsequent study decisions and supporting rationale.

- Non-assisted versus Assisted Takeoff: Assisted launch systems (e.g., rocket sled, electromagnetic sled, towed) on the scale necessary to meet the payload lift requirements are beyond the state-of-the-art for near-term application. Therefore, Non-assisted Takeoff was chosen.

- Vertical versus Horizontal Takeoff: Current horizontal takeoff vehicles and infrastructures are not capable of accommodating the gross takeoff weights of concepts needed to meet the payload lift requirements. Therefore, Vertical Takeoff was chosen.

- No Propellant Tanking versus Propellant Tanking During Ascent: Propellant tanking during vertical takeoff is precluded due to the short period of time spent in the atmosphere 1) to collect propellant or 2) to transfer propellant from another vehicle. Therefore, No Propellant Tanking was chosen.

- Rocket versus Air Breathing versus Rocket and Air Breathing: Air breathing and combined cycle (i.e., rocket and air breathing) propulsion systems are beyond the state-of-the-art for near-term application and likely cannot meet the lift requirements. Therefore, Rocket was chosen.

- Expendable versus Partially Reusable versus Fully Reusable: Fully reusable systems are not cost-effective for the low projected flight rates and large payloads. Near-term budget availability and the desire for a rapid development preclude fully reusable systems. Therefore, Expendable or Partially Reusable was chosen.

- Single-stage versus 2-Stage versus 3-Stage: Single-stage concepts on the scale necessary to meet the payload lift requirements are beyond the state-of-the-art for near-term application. Therefore, 2-Stage or 3-Stage was chosen.

- Clean-sheet versus Derivatives of Current Systems: Near-term budget availability and the desire for a rapid development preclude clean-sheet systems. Therefore, Derivatives of Current Systems was chosen.

E.3.2 The Use of a Trade Tree in "Launch Order, Launch Separation, and Loiter in the Constellation 1½-Launch Solution" [35]

The options that were evaluated in this decision analysis are depicted in the trade tree of Figure E-5. Two options were considered for launch order: launching Ares I first, followed by Ares V, identified as "I-V"; and launching Ares V first, followed by Ares I, identified as "V-I". In addition, two types of LEO loiter duration were considered. The first loiter option was to support only a single TLI window. The second loiter option was to support multiple TLI windows. Because of the limited loiter duration of the Orion crew module in LEO, the option to support multiple TLI windows is applicable only to a V-I launch order. Finally, options for the planned separation between the two launches of 90-minutes and 24-hours were evaluated. The ESAS baseline of a V-I launch order, a loiter duration that supports multiple TLI windows, and a launch separation of 24-hours is identified in Figure E-5.

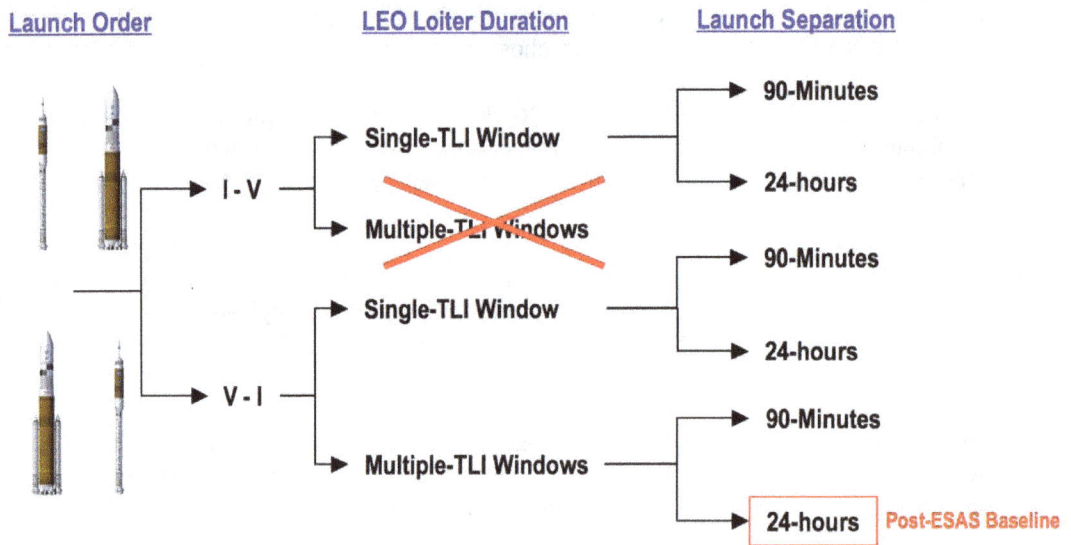

Figure E-5. Launch Order Analysis Trade Space

E.3.3 Hubble Space Telescope (HST) Servicing Analyses of Alternatives (AoA) [36]

Figure E-6 is the top-level robotic servicing decision tree that was used in the HST servicing analysis to scope out the space of mission concepts within which to compile specific alternatives. Due to a driving concern of an uncontrolled re-entry of HST, the expectation at NASA since program inception was that HST would be disposed of at end of life. Achieving NASA's casualty expectation standard of less than 1 in 10,000 requires some degree of active disposal of HST. Active disposal requires a minimum capability to rendezvous and dock with HST, and then either to boost the observatory to a disposal orbit or perform a controlled re-entry.

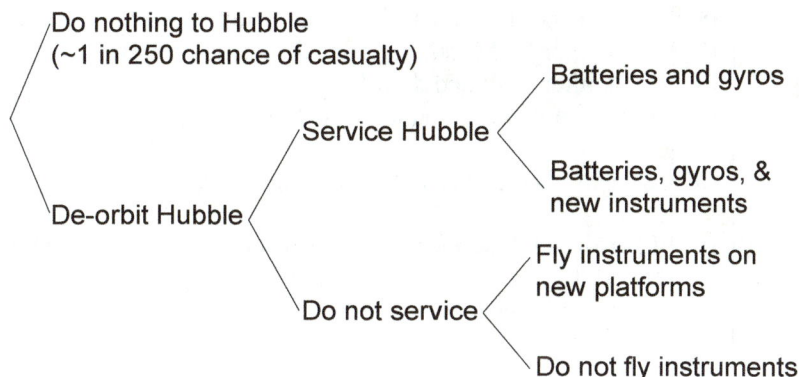

Figure E-6. Robotic Servicing Decision Tree

The alternative development began in brainstorming sessions in which the study team developed "clean-sheet" concepts, which encompassed doing nothing to HST, rehosting the SM4 instruments on new platforms, robotic servicing, and astronaut servicing. The brainstorming approach involved capturing the full set of ideas suggested by a large group of study team members during several alternatives development meetings. These ideas and concepts were then grouped into broad categories. The resulting database was augmented by internet and literature searches for related ideas, including the private sector responses on robotic servicing approaches and technologies.

Table E-2 provides the full set of alternatives, from which the final set was selected. A number of unorthodox options, including foreign participation in the development of a mission, delivery of ordnance and detonation just prior to the re-entry atmospheric interface, and forcing breakup at re-entry with a missile defense interceptor were dismissed as being too risky, as well as politically and/or technically infeasible.

Table E-2. Alternatives Brainstorming

Classification	ID	Description
Minimum Action	1	Do nothing
Minimum Action	2	Extend life through ground-based operational workarounds
Minimum Action	3	Use HST attitude modulation to control entry point
Rehost Option	4	Fly COS, WFC3, and FGS on new platform
Rehost Option	5	Fly COS, WFC3, and FGS replacement instruments on new platform
Rehost Option	6	Fly HST current and/or SM4 instruments on platforms already in development
Rehost Option	7	Fly HST current and/or SM4 replacement instruments on platforms already in development
Rehost Option	8	Rebuild HST
Rehost Option	9	Replace full HST capability on a new platform
De-orbit	10	Dock before EOM, de-orbit immediately

De-orbit	11	Dock before EOM, continue mission and de-orbit after EOM
De-orbit	12	Dock after EOM, but before EOL and de-orbit immediately
De-orbit	13	Dock after EOL and de-orbit
Service & De-orbit	14	Service-only mission (separate de-orbit mission)
Service & De-orbit	15	Launch SDM, service, stay attached for de-orbit
Service & De-orbit	16	Launch SDM, service, detach and station-keep, then reattach for de-orbit
Service & De-orbit	17	Launch SM and DM together, service, remove SM, DM stays on
Robotic Option	18	De-orbit vs. graveyard orbit
Robotic Option	19	Propulsive vs. attitude modulation reentry
Robotic Option	20	External attachment vs. internal replacement vs. internal attachment
Robotic Option	21	Autonomous vs. telerobotic docking
Robotic Option	22	Reboost as part of all servicing missions
Robotic Option	23	Timing of docking and de-orbiting (before/after EOM/EOL)
Robotic Option	24	Scope of servicing: full vs. extended life only (batteries + gyros + instruments vs. batteries + gyros)
Unconventional	25	De-orbit via foreign partners or commercial firms
Unconventional	26	Service via foreign partners or commercial firms
Unconventional	27	CEV crewed servicing mission
Unconventional	28	Uncontrolled de-orbit + KKV forced breakup at atmospheric interface
Unconventional	29	Uncontrolled de-orbit + detonation forced breakup at atmospheric interface
Unconventional	30	Use tug to deliver HST to ISS orbit for service by STS crew
Unconventional	31	Propulsively transfer to ISS and retrieve via shuttle
Unconventional	32	Point existing Earth-viewing orbital instruments upwards
Unconventional	33	Safe haven near HST for Shuttle servicing
Astronaut Servicing	34	Do SM4 (on STS)

A natural grouping emerged from this set of alternatives, whereby several complementary alternatives could be condensed into a single concept. Brainstorming was next linked with a deductive method. Options were distilled into four general families: Rehost, Disposal, Service, and Safe Haven as shown in Figure E-7.

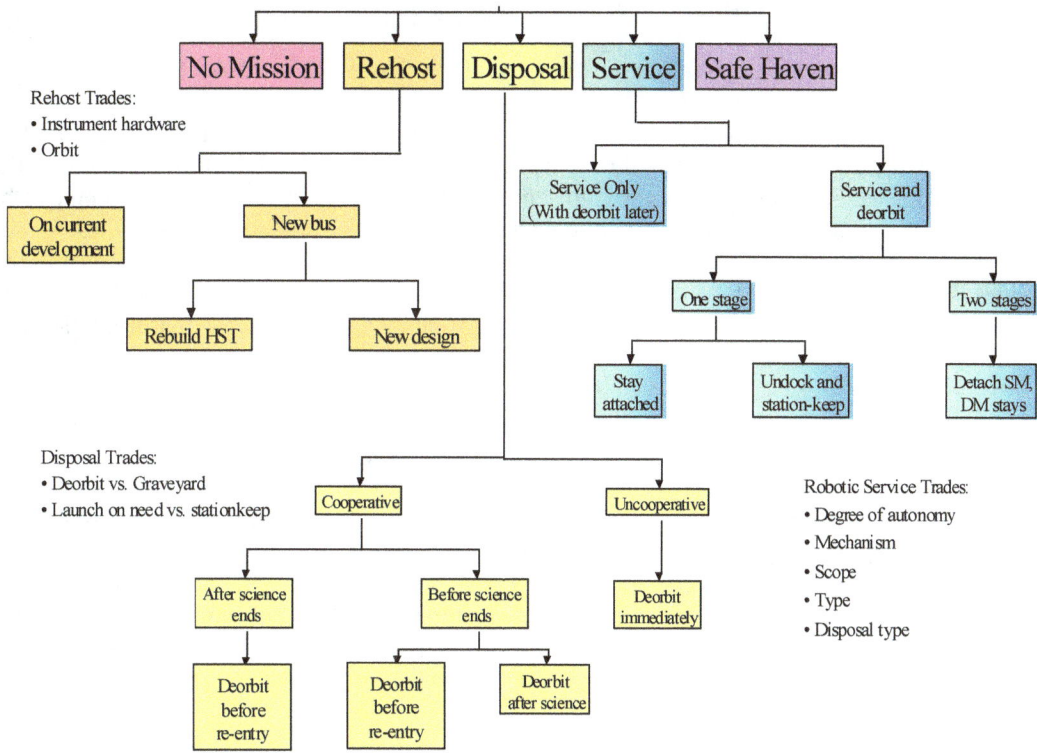

Figure E-7. Option Tree Analysis

The study team used a methodical approach to the process of selecting and constructing the final set of alternatives to ensure that the important elements in the robotic servicing trade space were included. The branches of the option tree were developed to a low enough level to cover various architectural and technology options at the conceptual level, such as the type of robotic mechanism and the amount of servicing performed.

E.4 Risk Analysis of Alternatives

E.4.1 Scenario-Based, Probabilistic Analysis of Performance in ESAS [23]

One of the issues analyzed in ESAS was that of monostability of the entry vehicle. In this context, a monostable entry vehicle will only aerodynamically trim in one attitude, such that the vehicle would always be properly oriented for entry (similar to Soyuz). Requiring an entry to be inherently monostable results in an outer mold line with weight and packaging issues. ESAS looked at how much benefit, from a crew safety risk (i.e., P(LOC)) standpoint, monostability provides, so that the costs can be traded within the system design. In addition, ESAS looked at additional entry vehicle systems that are required to realize the benefits of monostability and considered systems that could remove the need to be monostable.

The risk analysis consisted of two parts: a flight mechanics stability element and a P(LOC) assessment. The two pieces were combined to analyze the risk impact of entry vehicle stability. The risk assessment was performed using the simple event tree shown in Figure E-8, representing the pivotal events during the entry mission phase. Each pivotal event was assigned a success probability determined from historical reliability data. In addition, mitigations to key pivotal events were modeled using the results from the stability study, as were the success probabilities for ballistic entry. In the event tree, the "Perform Ballistic Entry" event mitigates the "Perform Entry" (attitude and control) event, while the "Land and Recover from Ballistic Entry" event replaces the "Land and Recover" event should a ballistic entry occur.

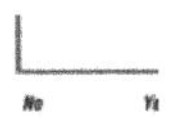

Figure E-8. ESAS Entry, Descent, and Landing Event Tree

E.4.2 Probabilistic Analysis of CLV Crew Safety Performance and Mission Success Performance in ESAS [23]

ESAS assessed more than 30 launch vehicle concepts to determine P(LOM) and P(LOC) estimates. Evaluations were based on preliminary vehicle descriptions that included propulsion elements and Shuttle-based launch vehicle subsystems. The P(LOM) and P(LOC) results for each of the CLV results are shown graphically in Figures E-9 and E-10, respectively. The results are expressed as probability distributions over the epistemic uncertainty modeled in the analysis, indicating the range of possible values for P(LOM) and P(LOC) given the state of knowledge at the time the analysis was done. Aleatory uncertainty has been accounted for in the analysis by expressing the results in terms of the probabilistic performance measures of P(LOM) and P(LOC). These measures represent the expected values for loss of mission and loss of crew, respectively.

Figure E-9. CLV LEO Launch Systems LOM

Figure E-10. CLV LEO Launch Systems LOC

E.4.3 Downselection in "Launch Order, Launch Separation, and Loiter in the Constellation 1½-Launch Solution" [35]

The launch order analysis down-selected to the preferred option through various down-select cycles that sequentially pruned options from the trade tree by focusing on various FOMs in each down-selection cycle until only one branch was left. Figure E-11 shows the overall trade tree and the down-selections made through 4 iterations considering various FOMs in pruning the tree, including a summary of each down-select rationale. The first down-select eliminating multiple TLI window support for a I-V launch order was based on the trade space constraint of not modifying the current Orion vehicle, which has a capacity to loiter in LEO for a maximum of four days, limited by consumables.

Opti⟨

Figure E-11. Launch Order Downselection and Rationale

E.4.4   Cost Sensitivity Study in "Launch Order, Launch Separation, and Loiter in the Constellation 1½-Launch Solution" [35]

The greatest level of uncertainty in the launch order analysis involves the transportation element replacement costs. Because of the uncertainty involving fixed and marginal costs for Altair and because of the uncertain nature of cost estimates in early design, large shifts could potentially occur in the cost data for all elements as the designs mature.

Figure E-12 was used to determine the optimal launch order for any set of Ares I and Ares V stack replacement costs, based on minimizing the expected cost of failure. The cost of the Ares V stack is specified on the horizontal axis, and the cost of the Ares I stack is specified on the vertical axis. The sloping red line in the center of the figure represents the break-even cost boundary. If the set of costs is below this line in the light-blue region, then the I-V launch order is preferable. If the set of costs are above the red line in the light-green region, then the V-I launch order is preferable.

The intent of the figure was to provide a visual indication of how much change could occur in the cost estimates before the launch order decision would be reconsidered. The analysis cost estimates are represented as a horizontal bar on the chart. The Ares I cost is normalized to 1. The

Ares V cost is represented as a range of 1.65 to 2.88 times the cost of Ares I, which represents the full range that is produced by the possible inclusion of fixed costs. It is evident that, even at the low end of the Ares V cost range, a large margin still exists before the break-even point is reached.

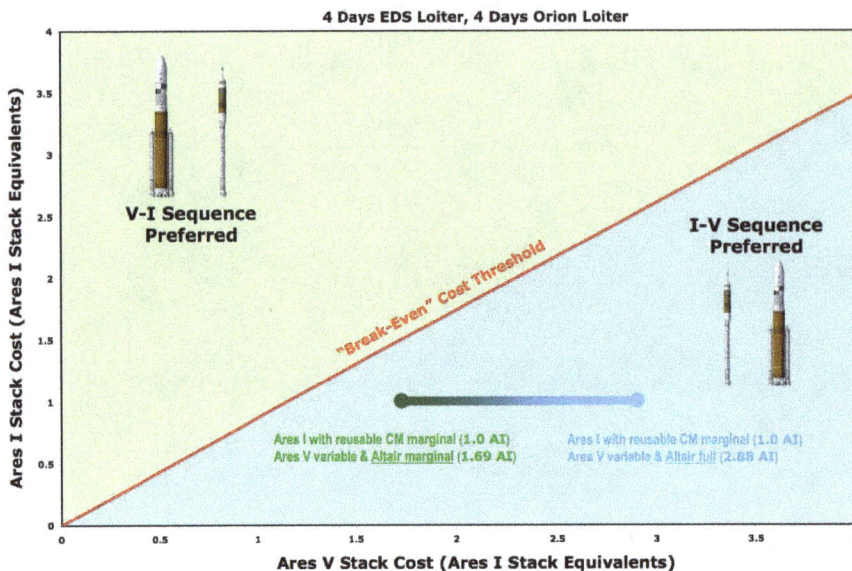

Figure E-12. Launch Decision Relative to Ares I and Ares V Stack Costs

E.4.5   Performance Communication in ESAS [23]

A summary of the ESAS FOM assessments for the Shuttle-derived CLV candidate vehicles is presented in Figure E-13. The assessment was conducted as a consensus of discipline experts and does not use weighting factors or numerical scoring, but rather a judgment of high/medium/low (green/yellow/red) factors, with high (green) being the most favorable and low (red) being the least favorable.

The Shuttle-derived options were assigned favorable (green) ratings in the preponderance of the FOMs, primarily due to the extensive use of hardware from an existing crewed launch system, the capability to use existing facilities with modest modifications, and the extensive flight and test database of critical systems—particularly the RSRB and SSME. The introduction of a new upper stage engine and a five-segment RSRB variant in LV 15 increased the DDT&E cost sufficiently to warrant an unfavorable (red) rating. The five-segment/J–2S+ CLV (LV 16) shares the DDT&E impact of the five-segment booster, but design heritage for the J–2S+ and the RSRB resulted in a more favorable risk rating.

Applicability to lunar missions was seen as favorable (green), with each Shuttle-derived CLV capable of delivering the crew to the 28.5-deg LEO exploration assembly orbit. Extensibility to commercial and DoD missions was also judged favorably (green), with the Shuttle-derived CLV providing a LEO payload capability in the same class as the current EELV heavy-lift vehicles.

Figure E-13. Shuttle-Derived CLV FOM Assessment Summary

E.4.6   Performance Communication in the Hubble Space Telescope (HST) Servicing Analyses of Alternatives (AoA) [36]

In order to communicate the cost-effectiveness of each alternative, several primary MOEs were combined into one governing metric. To develop this metric, an expected value approach was taken. Expected value theory is based on the notion that the true, realized value of an event is its inherent value times the probability that the event will occur.

The expected value approach took into account the performance of each alternative relative to post-SM4[33] capability (MOE #5), the probability of mission success (MOE #4) and the probability that the HST will have survived to be in the desired state for the mission (MOE #3), which is a function of HST system reliability and development time (MOE # 2). The calculation of expected value was the value of the alternative times the probability of the alternative successfully completing its mission:

$$\text{Expected Value} = \text{MOE \#3} * \text{MOE \#4} * \text{MOE \#5}$$

Figure E-14 illustrates the results of the combined expected value plotted against life-cycle cost. The results indicate that the disposal alternatives provided no value relative to observatory capability. The expected value calculation also indicated that rehosting both the SM4 instruments on new platforms provided higher value at equivalent cost to the robotic-servicing missions. There was, however, a gap in science with the rehost alternatives that was not captured in this expected value calculation.

---

[33] Servicing Mission 4 was the HST servicing mission previously scheduled for 2005.

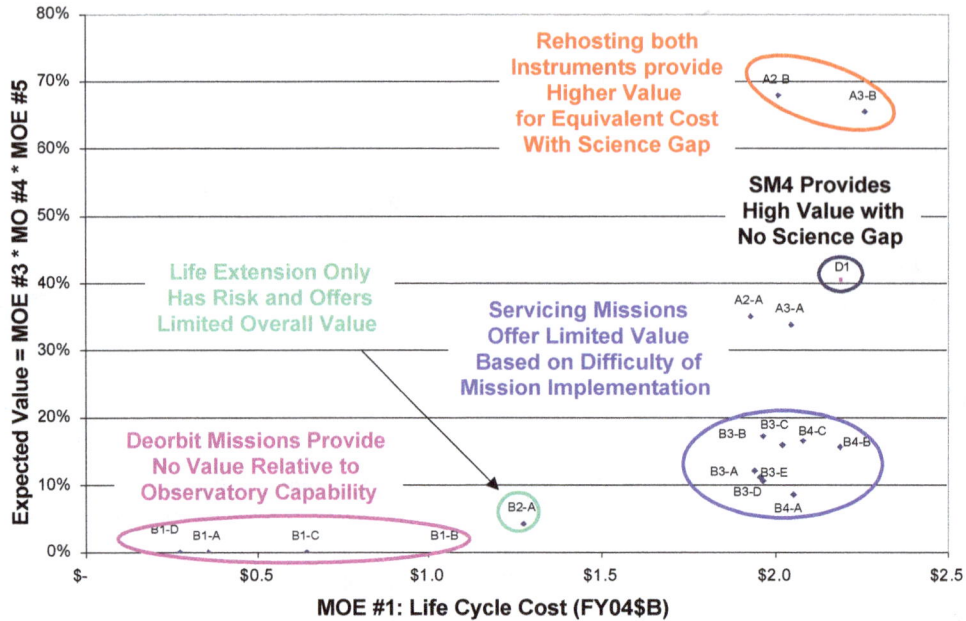

Figure E-14. Expected Value versus Life Cycle Cost

The robotic servicing alternatives cluster in the lower right corner of the plot, suggesting that the value of these alternatives was limited based on difficulty of the mission implementation, the complexity of the servicing mission, and the reliability of HST after servicing.

SM4 had costs in the same range as the rehost and robotic-servicing alternatives. It had the added benefit of higher probability of mission success than the robotic servicing missions, and did not suffer from the gap in science associated with the rehost alternatives.

APPENDIX F: PRACTICAL ASPECTS OF THE RISK MANAGEMENT PLAN

The following subsections briefly discuss some of the practical issues spanning the entire RM process that must be addressed by the RM plan.

F.1 Staffing Resources

The project manager will usually assign a risk manager with responsibility for coordinating and overseeing the implementation of the RM process. For small projects, this may be a quarter- to half-time job. For larger projects this will be a full-time job and may have to be directly supported by additional individuals. For example, some projects may use a risk integrated product team (IPT) to oversee the project RM process.

It is important to clearly establish roles and responsibilities for risk owners and other participants in the RM process. Risk owners who have ownership of individual risk items and handling plan implementation directly support the risk process with some pre-established minimum level of commitment (in terms of hours per week or per month). This support includes providing periodic risk updates to the risk manager and/or risk IPT, briefing the risk board on current progress, and receiving / providing input at contractor RM activities.

A similar level of commitment is also required from project management personnel, not including time spent in implementing any selected risk mitigation strategy, since these activities are part of the risk owner's and/or project manager's normal duties. As warranted by the nature of the issue, assessment activities for some risk items may also required additional effort by the risk owner or by the technical personnel to whom assessment has been delegated.

The establishment and coordination of a risk management board and the ground rules by which it is to be run, including the expected frequency of sessions to be held, is also part of the points addressed by the RM plan. For typical projects, the frequency of RM Board meetings varies from the a monthly to a quarterly basis.

F.2 Plan Updates

The RM plan is reviewed and updated as the project moves through each key decision point. The following areas are typically to be reviewed and adjusted as needed:

- Assessment level of detail and focus (risk tolerance changes as the program progresses)

- Review cycle/board frequency (as the system moves into production, the frequency must often increase so that the risk process can support more fast-paced project decisions)

- Reporting formats and RM tool capabilities

- Interaction with contractor and lower-level / higher-level risk processes

F.3 Training

All personnel in a project should receive basic RM training that includes an overview of terminology, RMt process flow, roles and responsibilities, and instruction in the identification, assessment, and mitigation-planning of project risks. For larger projects, several levels of training may be necessary:

- All personnel: basic RM training

- Risk IPT members: detailed process and tool training

- Program management: process overview and reporting formats

- Contractors and external organizations: expectations for interaction with the NASA process

Retraining may be necessary if the RM plan is changed or deficiencies are found during process implementation.

F.4 Interaction of the Project and Contractor Processes

A project tracks and manages risks from the Government perspective and monitors contractor RM efforts. The NASA RM process selectively utilizes information and data generated by the contractors' RM activities and addresses risks from a Government perspective.

Each level of the RM effort may use different risk scales and assessment criteria tailored to the magnitude of risk consequence and the risk tolerance at that level. Clear definitions of assessment scales at each level, conversions from one scale to another (if necessary), and elevation criteria are required and should be explicitly addressed. Elevation criteria describe when a lower-level risk should be brought up to the next level in the RM process. These criteria may be program-specific and dependent on the acquisition structure of the program. In general, however, risks may be elevated from the contractors' side of the process and reassessed on the Government side when action and tracking by NASA are required:

- Mission or project risks rated as "red" or "mission critical" on the contractor's scale

- Items that are the Government's responsibility to control (e.g. GFE)

- Contractor items that are caused by or may impact other portions of the project out of the control of the contractor

- Contractor items whose handling options may have a cost impact for the Government

- External interface issues

- External risks (e.g., funding instability)

When a project is divided into multiple, relatively independent segments, and/or is executed by more than one prime contractor, a multi-level RM process may be necessary both to oversee each contractor and to roll-up risk information to a system level. Ideally, this type of process permits the elevation of issues to the level where the most effective decision-making process can take place and the appropriate technical and financial resources are available for their successful handling. At the same time, this process should facilitate vertical awareness and independent review to keep underestimation, overestimation, or mishandling of issues to a minimum (see [37] for more detailed discussion of some of these issues).

APPENDIX G: HYPOTHETICAL INDIVIDUAL RISKS USED FOR THE PLANETARY SCIENCE EXAMPLE IN THE CRM DEVELOPMENT

No. & Title:

1(a). Planetary Contamination

Risk Statement:

Given that [CONDITION: the state of knowledge of Planet X's atmosphere is limited; the fact that it is difficult to ascertain more information about Planet X's atmosphere from Earth; and the fact that the spacecraft contains radioactive material], there is a possibility of [DEPARTURE: unanticipated atmospheric characteristics during the aerocapture maneuver at Planet X leading to a less-than-optimal trajectory] adversely impacting [ASSET: the spacecraft], thereby resulting in [CONSEQUENCE: spacecraft breakup and radioactive contamination of Planet X]

Narrative Description:

The atmosphere of Planet X has been observed with ground-based and Earth-orbital telescopes at various times, including during eclipses, and spectral analysis of the data has been performed. There have also been flybys to observe atmosphere thickness, species, and density. Uncertainties in the results are large because of inherent variability in the atmosphere, both spatially and timewise, making it difficult to make global inferences from limited observations. Other direct sources of uncertainty include limitations in instrument accuracy and variable solar flux effects. Additionally, there is considerable inherent uncertainty in the models used for calculating thermal responses and stresses in the heat shield, bond, and structure because these models are based on assumptions about the effects of ionizing radiation on heat transfer and the condition of the vehicle surface as it affects boundary layer transition. Associated testing in wind tunnels, hot gas facilities, and plasma arcs may be of limited applicability for the Planet X atmosphere. If the spacecraft should break up during the aerocapture maneuver, analysis shows that because of the extra-orbital velocity of the spacecraft (hyperbolic entry), it is highly likely that Pu would scatter and some fraction would reach the surface.

No. & Title:

1(b). RCS Damage

Risk Statement:

Given that [CONDITION: the state of knowledge of Planet X's atmosphere is limited; the fact that it is difficult to ascertain more information about Planet X's atmosphere from Earth; and the reaction control systems fielded to date has not needed to operate in such harsh (hyperbolic entry) environments], there is a possibility of [DEPARTURE: unanticipated atmospheric characteristics during the aerocapture maneuver at Planet X] adversely impacting [ASSET: the exposed RCS components], thereby leading to [CONSEQUENCE: damage to the RCS system making it unable to perform orbital maneuvers and achieve a circular orbit].

Narrative Description:

The RCS is exposed to the aerocapture environment because of the need for it to operate before and after the aerocapture maneuver (as well as possibly during the maneuver). There are no data for RCS behavior in the hyperbolic entry conditions that would be encountered at Planet X, as the RCS components have never before been required to operate after enduring the environments of an aerocapture maneuver. Hence certain parts of the RCS such as the motors, valves, and sensors could be operating in a regime for which they are not suited. Ground tests are planned for the RCS under conditions simulating hyperbolic planetary entry, but the conditions it will be exposed to during actual flight are highly uncertain. Similar risks pertain to other assets that are exposed to the environment, such as navigation sensors and communication antennas (but not science sensors since these are assumed to be shielded).

No. & Title:

2: Pu^{238} Availability

Risk Statement:

Given that [CONDITION: the supply of available Pu238 is limited for deep space missions; and Congress has not approved funds for processing Pu in the United States], there is a possibility that [DEPARTURE: the cost of Pu may increase drastically] impacting [ASSET: the spacecraft electrical power supply], thereby leading to [CONSEQUENCE: inability to perform the mission within the required cost].

Narrative Description:

The current assessment in the Agency's Pilot Risk List is that since the US has discontinued production of Pu238, and Congress has not approved funding for production in FY11, and Russia will no longer supply Pu238 without a new contract, there is a possibility that there will not be enough Pu238 for some future deep space missions. It may be possible to conserve on Pu (and to test out a new technology) by using Advanced Stirling Radioisotope Generators (ASRGs) in place of the more traditional MMRTGs to provide power. The ASRGs require much less Pu than the MMRTGs for their operation, but their reliability for long operating times (12 year mission life) is not known.

No. & Title:

3: Thrust Oscillations

Risk Statement:

Given that [CONDITION: development testing indicates there may be thrust oscillations in the launch vehicle during ascent that have not been accounted for in the design of the scientific instrumentation in the payload and that these oscillations may be near the resonance frequency for the payload], there is a possibility of [DEPARTURE: operational vibrational stresses exceeding design limits] adversely impacting [ASSET: the scientific instrumentation], thereby leading to [CONSEQUENCE: degradation or failure of the ability to obtain and transmit scientific data].

Narrative Description:

Developmental testing showed that vibrations produced by the first stage may align with structural frequencies in the axial stack. Because the science package designers were aware that thrust oscillations had occurred occasionally (but rarely) in the testing program for an Earth orbit mission, they performed some preliminary analysis of the potential effects of thrust oscillation for the planetary exploration mission when they saw the data from the first stage testing. Their results indicated that loading of the science package from first stage oscillations could be significant, compounded by the fact that the present vehicle has a particularly high proportion of payload mass to total mass. The high payload mass fraction results from the fact that the spacecraft does not have much propellant. Normally, the propellant would act as a damper for vibrational excitations. The uncertainty in the coupling between first stage excitations and the science package is complicated by the fact that there is no integrated mechanical model for the stack. An integrated mechanical model will be developed, validated, and exercised to reduce that source of uncertainty in time for CDR.

No. & Title:

4(a): DMS Institutional Risk

Risk Statement:

Given that [CONDITION: the current Document Management System utilizes a commercial database querying utility, and that the company that provides the utility has indicated they will no longer support and maintain it], there is a possibility that the software for the utility will either reach a failed and nonrepairable state or become obsolete] adversely impacting [ASSET: the Document Management System], thereby leading to [CONSEQUENCE: inability to meet data management needs of the Center].

Narrative Description:

The Data Management Support organization has the requirement of maintaining an electronic DMS that meets the needs of the Center. The electronic system must be robust enough and efficient enough to support the communication needs of all the projects in the Center, including the receipt, cataloging, and dissemination of Agency documents, external documents that pertain to the Center's projects and functions, and contractor deliverables. This requirement is put in jeopardy by the fact that the database querying utility that supports the DMS is provided by a commercial vendor who has indicated they will be discontinuing their support and maintenance of it due to financial pressures. If the utility should fail or become obsolete, it may not be possible to keep the DMS functional at the level required.

No. & Title:
   4(b): DMS Project Risk

Risk Statement:
   Given that [CONDITION: the current Document Management System utilizes a commercial database querying utility, the company that provides the utility has indicated they will no longer support and maintain it], there is a possibility that [DEPARTURE: the software for the utility will either reach a failed and nonrepairable state or become obsolete] adversely impacting [ASSET: communications within the Project], thereby leading to [CONSEQUENCE: a delay in the launch date].

Narrative Description:
   If the DMS fails or becomes obsolete, the project will have to rely more heavily on other existing communication venues such as meetings, emails, and interoffice mailings until a new DMS can be implemented. The resulting slowdown in communication will affect the delivery dates of several assets as well as lengthen the time required for system integration. In addition, there will be an increase in the probability for errors that are related to inadequate communication.

No. & Title:
   5: Valve Effect on Mass Margin

Risk Statement:
   Given that [CONDITION: a type of valve in the RCS has failed due to excessive corrosion during a qual test and needs to be replaced], there is a possibility that [DEPARTURE: it may be necessary to replace it with a valve type that is significantly heavier than the current choice] adversely impacting [ASSET: the RCS], thereby leading to [CONSEQUENCE: failure to meet the mass requirement for the RCS and possibly for the entire spacecraft].

Narrative Description:
   During qualification testing, valve Z failed due to the corrosive effect of certain gaseous chemical species on the valve seals and to a lesser extent the valve body. These species are much more prevalent in the Planet X atmosphere/exosphere than in other planetary atmospheres/exospheres, which explains why similar valves have not failed during qual testing in other projects. Replacement valves of similar or lesser mass may also be susceptible to corrosive failures. If it becomes necessary to use a valve that is significantly heavier than the current choice, the additional mass could result in a significant reduction in the mass margin for the RCS. The effect on the mass margin for the entire spacecraft may also be of concern.

No. & Title:
6: Sensor TRL

Risk Statement:
Given that [CONDITION: the unique atmospheric sensors needed for this mission are at TRL 2 and must reach TRL 8 prior to integration at 40 months], there is a possibility of [DEPARTURE: the sensors being unavailable or late] adversely impacting [ASSET: the science package], thereby leading to [CONSEQUENCE: failure to meet the 40-month delivery date for the science package integration into the spacecraft bus].

Narrative Description:
In order to get a relatively complete picture of Planet X's atmosphere's fluctuations due to the solar wind, it will be necessary for the spacecraft to provide atmospheric sensor data for a minimum of 6 Earth-months. A new sensor is needed to capture information about Planet X's atmosphere because of the fact that it is unique compared to the other planets. Presently the sensor has been demonstrated only in a lab environment, placing its technology readiness level at TRL 2. In addition the substrate for the sensors has not been demonstrated to be manufacturable. The new technology is part of a low-fidelity science package in order to keep the payload weight down. The planned launch in 55 months to take advantage of the solar maximum cycle must allow sufficient time for technology development and sensor manufacturing, testing, and integration.

No. & Title:
7: Video Sensor Stray Light

Risk Statement:
Given that [CONDITION: another mission experienced degraded video clarity due to excessive light entering the science package], there is a possibility of [DEPARTURE: excessive light entering the science package during this mission] adversely impacting [ASSET: the video system], thereby leading to [CONSEQUENCE: an unacceptable loss of data quality].

Narrative Description:
During NASA's Mission M, the video images transmitted to Earth were degraded by unwanted light intrusion into the camera box. What is known from the video is the direction from which unwanted light entered the camera lens. What is unknown is whether the problem was caused by direct sunlight entering through a small unknown aperture in the casing of the science package, by reflection of sunlight off an external antenna, or by a light source within the spacecraft. The camera and antenna configurations for both missions are similar. Although the amount of overexposure was not enough to jeopardize the mission objectives for Mission M, the degradation of the images might be worse if the same problem occurred for this mission.

No. & Title:

8: Staffing for Legacy Software

Risk Statement:

Given that [CONDITION: the decision has made to adapt legacy software for this mission and the Agency projects a scarcity of qualified programmers familiar with the legacy language], there is a possibility that [DEPARTURE: there may be insufficient staffing at the high labor categories] adversely impacting [ASSET: the control software] which could result in [CONSEQUENCE: delays in the delivery of the software and/or software reliability issues].

Narrative Description

NASA's budget for future years reflects reduced funding for certain legacy programs and in some cases outright termination of the program. It is anticipated that this will lead to retirements and/or resignations as qualified staff review their options. In particular, it is possible that a predominance of the staff departures may be in the higher labor categories among people who have experience using the programming languages associated with legacy software.